ETHICS IN TECHNICAL
COMMUNICATION

ETHICS IN TECHNICAL COMMUNICATION

PAUL M. DOMBROWSKI

ALLYN AND BACON

Boston ■ London ■ Toronto ■ Sydney ■ Tokyo ■ Singapore

Vice president, humanities: *Joseph Opiela*
Series editorial assistant: *Mary Beth Varney*
Executive marketing manager: *Lisa Kimball*
Composition and prepress buyer: *Linda Cox*
Manufacturing buyer: *Suzanne Lareau*
Cover administrator: *Jenny Hart*
Editorial-production service: *Shepherd, Inc.*
Electronic composition: *Shepherd, Inc.*

Some documents in this text have not been reproduced directly, but have been reset to match the originals as closely as possible.

Dombrowski, P. M. (Paul M.)
    Ethics in technical communication / Paul M. Dombrowski.
        p.   cm.
    Includes bibliographical references.
    ISBN 0–205–27462–5
    1. Communication of technical information--Moral and ethical
aspects. I. Title.
T10.5.D66 1999
174'.96--dc21   99–36590
                                                                              CIP

Printed in the United States of America

10 9 8 7 6 5 4 3 2 1   03 02 01 00 99

CONTENTS

CHAPTER SEVEN

Star Wars: Hope vs. Reality   190

PART III   APPLICATIONS: HYPOTHETICAL CASES

CHAPTER EIGHT

Ethics Exercises   233

PREFACE

Ethics is always involved in technical communication, though only in the last twenty or so years has it become an important topic in technical communication publications. In the early years of technical communication as a profession, ethics was important, though it was not spoken of as such. Good technical communication was expected to be clear, lean, and accurate, as it still is. Technical communicators were expected to practice their art so as to exemplify these values. Thus characteristics of style and of stance toward the subject matter and toward the audience reflected a system of values, a work ethic of technical communication, if you will.

The face of our field has changed considerably since those early years. Though the transmission of factual technical information still lies at the core of technical communication, we now realize that it is more than that. There are successive layers of functions and purposes of technical discourse extending outward from that core. We understand that technical communication is not just about technical information but also about how it is used. How that technical information will be received by a very human audience, what function it will serve, and how it will be put to practical effect are also important dimensions of technical communication apart from the information itself. How all this can be done as efficiently as possible and without costly mistakes are important issues too. How we address such issues determines how good our technical communication is, its comparative value in relation to other ways of communicating the same information. And ethics is all about values, about what is right and good.

We have also come to appreciate the complex role of social context in creating, shaping, and giving meaning to information, as well as in giving direction to its use. Just as a screwdriver is not simply a thing but a technological artifact that has meaning and use only in the context of a culture of screws and other hardware, any technology is intertwined in its social context in many different ways. It not only reflects that context but even makes it possible in the first place. This social context extends outward indefinitely for whatever technical artifact we are considering, limited only by the constraints we choose to set on our considering. In our screwdriver example, this includes the entire social fabric of resources, economics, even politics that establishes such machines as being important, desirable, and valuable. An Amish community, for example, has no use for any electronic gadgets held together by our screws, even in a simple device like a lightning rod. This context, the Amish community, consists of people and society, and ethics is all about values involved in our human relations.

We have come to appreciate as well the complex and powerful role of language in forming this social context. Language use defines social discourse communities and corporate cultures, articulating what is important and valuable as well as what is not. The merger of Chrysler Corporation in America and Daimler Benz in

Germany in 1998, for example, caused these corporations to become more conscious of their language and of the values reflected in language use. Their discourse is now carefully crafted to bind the two into a workable single entity with a common culture and language with shared values, a shared ethic.

Thus we have come to a modern appreciation of an ancient awareness, namely that ethics and rhetoric as communication are not just allied fields but two sides of the same coin: you cannot have one without the other. In classical times, many rhetoricians recognized that discourse embodies values. Even though Aristotle excluded scientific and mathematical knowledge from the realm of rhetoric, he was acutely aware that rhetoric was frequently about achieving ends, which are defined by values and are made possible by technical means. In our own times, communication and rhetoric scholars have emphasized the inescapable interconnection between discourse and values or ethics, including scholars as otherwise disparate as Richard Weaver and Michel Foucault. Feminist critics, culture critics, and others have caused us to examine the role of technology in our culture in many different ways and have made us more conscious of technical discourse in making that role possible.

Also in recent times, traditional ethicists such as Langdon Winner, Jacques Ellul, and Leo Marx have turned their attention to technology itself. They have made the public more ethically sensitive to the role of technology in our lives, whether in controlling air pollution by insisting on cleaner engines or in challenging the use of radioactive generators in space probes. Since technology itself has ethical significance, naturally, technical communication as discourse about the use of technology has ethical significance as well.

In all these ways and more, ethics is involved in technical communication. In this book the term "ethics" includes the field of study about judging the more desirable from the less, about particular theories about such judgments, and about the values and systems of values which guide or reflect such judgments. Ethics also includes both personal and social judgments, though usually excluding the law. Ethics is concerned here with what is problematic and cannot be assumed because the need to weigh and deliberate, whether interpersonally or intrapersonally, defines the ethical dilemma.

Throughout this book, we will be discussing ethics in this broad sense, including not only the act of communication or its mechanical correctness, important though these are, but also where technical information came from and how it likely will be used. We will also discuss the impression technical information leaves on its audiences including impressions about values, uses, and goals. This broad-stroke approach to ethics encompasses as much territory as possible within our ethical responsibilities as technical communicators. This wide breadth is not meant to suggest that we relieve any other parties to our communication transactions of their own responsibilities, such as the subject matter experts or the audience. It only suggests that our responsibilities extend beyond the bare relaying of information between expert and audience, even though that act has its own ethical significance.

AUDIENCE

This book is intended for advanced undergraduate and graduate audiences. It assumes that you have some familiarity with technical communication already including the theory that explains, derives from, and guides our craft. Though it does not assume any formal exposure to ethics, you do need to be willing to consciously reflect on the subject of ethics as it relates to you as well as to others, to entertain a range of possible legitimate courses of action, and to make an effort to articulate why or how one course of action might be ethically preferable to others. Though ethics is primarily (though not always solely) a personal matter, you will be expected to discuss ethics and weigh particular cases with others in the class in a way typical of college or university advanced level courses. You will need to be willing to listen to differing opinions with the same respect that you wish your own opinions to be heard, not that you necessarily will agree with them but that you will agree that they have a right to be heard and recognized. You will also need to articulate your own ethical comments as clearly as you can, making your points and explaining them with supporting comments that will make sense to others, though they may not agree. Because communication communicates values as well as other content, our communications *are* to some extent our values and vice versa. For that reason, what we say, as a form of what we do, is what ethics is all about.

PURPOSE

The purpose of this textbook is to get you thinking and talking and doing ethics as a part of your professional life. It aims to stimulate class discussion and reflection about and to increase your awareness of the ethical dimension of technical communication. You will be introduced to several different ethical theories or perspectives. For the most part, no one particular ethical theory will be urged over any others, nor any one particular direction of judgment over others, though we will see that most ethical theories are in general agreement in their application to most cases.

   This textbook does not take an absolutist stance that assumes that there is only one truly ethical course of action. On the other hand, it does not take a strong relativist stance either, which would assume that there are many ethical stances all of which are equally valid and preferable, which would leave us without grounds for choosing all or none of the above; that would make ethics arbitrary. After all, since ethics for you is all about what you would do, we assume that you will do something in a given case; you cannot do everything and cannot do nothing. What we ask of you is that you try to make clear to yourself as best you can, and express and explain to others, what that something is.

ORGANIZATION

This textbook has three major sections. The first section covers the theory and history of ethics. Chapter One explains what ethics is for the purposes of this book. Chapter Two reveals the connection between ethics and communication or rhetoric. Chapter Three summarizes some of the major ethical theories: Aristotle's ethics of virtue and character, Kant's ethics of duty, utilitarianism's ethics of calculated benefits weighed against costs, feminist ethics opposing gender bias in technology, science, and discourse, and a contemporary ethics of care.

The second section presents four real, complex technical communication situations in detail and applies the major ethical theories to them. Chapter Four deals with supposed "technical" and "scientific" information from the Nazi regime, examining the origins and uses of this information. It also discusses controversies in the U.S. Environmental Protection Agency in the late 1990s and revelations by the Clinton administration of human radiation experiments in the United States since the 1940s. Chapter Five deals with the space shuttle *Challenger* disaster in 1986, focusing on differences in language use that reflect crucial assumptions and perspectives in key technical documents. It also contrasts the *Challenger* documents to key technical documents relating to the Three Mile Island nuclear disaster in 1979. Chapter Six deals with the U.S. tobacco industry and its representation of crucial technical and scientific information in key documents from the 1950s to the late 1990s. Chapter Seven deals with Stars Wars, the common name for the Strategic Defense Initiative ballistic missile defense program, focusing on exaggerated and unrealistic claims represented as factual information. It also summarizes the earnest ethical debate among scientists and academics about this high technology.

The third section consists of a single chapter that presents six hypothetical cases dealing with ethical issues in realistic but fictional technical communication situations. In several cases, slight variations on the basic case are made in order to clarify the importance of particular values in how you make your judgments (a common values clarification technique). Possible ethical appraisals of these hypothetical cases are sketched at the end of the chapter.

Each chapter involving cases includes a detailed ethical appraisal of the material focused on in the chapter. Also included at the end of each chapter is a set of topics for class discussions or for course papers, and Web sites for research and additional topics.

FEATURES

This book is a solid blend of theory and practice. Nearly one half of the text is devoted to the principles, perspectives, theory, and history that are the prerequisites of any thorough discussion of ethics in technical communication. The second half is devoted to the practical application of the ethical awareness cultivated in the first half. Four real instances show the many facets of any practical ethical situation, and the very human dimensions of any ethical dilemma. Though few of us will ever be involved in such momentous events, each of us will likely face situations of greater ethical complexity and depth than we might prefer but which we must recognize and accept. Six hypothetical cases afford the opportunity to explore ethical territory on your own, whether individually or in groups. Several variations of these cases allow you to clarify the values at work in your ethical judgments.

ACKNOWLEDGMENTS

Many people have made this book possible. I am especially appreciative of Sam Dragga at Texas Tech for inviting me to write this book. Sam Dragga also invited me to serve as the first chair of the Ethics Committee of the Association of Teachers of Technical Writing and to coordinate drafting our code of ethics. This experience afforded me a fuller appreciation of ethics in practical application. Joe Opiela at Allyn & Bacon provided both patient support and editorial guidance. Several diligent reviewers provided further guidance: Susan Booker, Iowa State University; Melody Bowdon, University of Arizona; and TyAnna Herrington, Georgia Institute of Technology. The support of my family throughout this long process is both the first and last link in the chain of my appreciation for making this book possible.

CHAPTER 1

NATURE
OF ETHICS

Ethics in the field of technical communication is an increasingly important topic. Most new textbooks incorporate throughout the text or in a separate section a treatment of ethics. Nearly all major technical communication conferences offer panels on ethics. Almost all technical communication journals have published articles about ethics, and many have published special issues devoted only to ethics, including *Technical Communication Quarterly, Technical Communication,* and *IEEE Transactions on Professional Communication.* University and college curricula are offering required or elective courses in ethics in technical communication. Practically all professional and academic organizations involved with technical communication have developed guidelines or codes of ethical conduct.

The driving force behind these scholarly and professional developments is a broad new awareness among technical communicators of the ethical implications of our work. This awareness is due in part to the recognition of many important ethical lapses in recent years involving communications about technology. The major technological disasters of recent years seem to have been linked in various ways to problems in communication in this country—from the danger of charred O-rings on the *Challenger* to the danger from leaking silicone breast implants. In other countries we have learned of inadequate safety documentation at Bhopal and botched technical procedures at Chernobyl.

This growing interest within our profession resonates with the rising concern among the general public about ethics in nearly every facet of our lives. We hear about public outcries for more ethical conduct regarding everything from financial contributions for political campaigns to the reporting of scientific research. Increasingly, the public and the news media have deliberately begun investigating and critically examining the ethics of all manner of activities. The public and the media that inform them are assuming the responsibility of deliberating about ethics in situations that previously have been left to others.

Ethics involves making judgments about values. In business and industry, citizens have scrutinized and criticized on ethical grounds the values underlying some of the nation's largest institutions. The deregulation of a whole host of industries from transportation to public utilities, for example, has occurred in part as a response to an underlying value: the supposed desirability of a free-market economy above most other considerations. Others, however, have argued the importance of other values, such as public safety in the airline industry, which, they contend, outweigh the desirability of an unregulated market.

Even regarding technology and science themselves, the public is increasingly active in scrutinizing research, expenditures, and goals. Values are now shifting in many different ways. The specialized nature of knowledge gained from technology and science is no longer seen as an adequate warrant for excluding the public from policy decisions. Langdon Winner and Craig Waddell, for example, have examined the new role of the public in decision making about technology and science policy. Increasingly, the public is unwilling to allow scientists to decide unilaterally the desirability of scientific programs and activities. The public is involved in discussions and decision making about the safety of radioactive waste disposal, for example. In previous decades national security and scientific progress were valued over the interests of particular citizens or regions of the country. Ethical concerns are raised even about space. Recently a number of groups protested the launch of the *Cassini* space probe because it utilizes plutonium as its energy source, a chemical of uncommonly powerful radioactivity and toxicity.

Similarly in the technology of communication itself, concerns are repeatedly and vocally raised about values. One of the chief merits of the World Wide Web as a communications medium, supporters explain, is its innate democratizing influence in treating all users equally. It is owned by no one and by everyone, and its control and usage are decentralized, so it empowers all users equally. Others, however, argue that it selectively and unfairly facilitates communications among those already advantaged and so aggravates rather than ameliorates social inequalities. Similarly, the breakup of the perceived monopoly on communication by Ma Bell was justified on the grounds of the value of unrestricted free trade and fair competition, not on the grounds of technical excellence or efficiency. Values, we see, shape communication technology itself, imparting an ethical cast to such technological developments.

In information technology, ethical concerns are raised about issues of privacy, ownership of information, copyright, access, freedom of speech, personal and national security, and access to markets in other countries—all matters of values with powerful implications for how these new technologies will be used. Arguments over who gets what share of the bandwidth are presented partly on the basis of the general good to the public, a civic value.

Returning to our own field of technical communication, the overt interest in ethics has until recently perhaps been less pronounced than in other fields. The reasons for this are many, but one of the most important is the nature of technical communication itself and its history. As many writers in our field have pointed out, historically technical communication has been understood in a fairly mechanistic way. It has been viewed as the articulation and dissemination of information about technology—the technology and information being assumed givens. From this point of view, the role of technical communication was chiefly to relay information to the recipient as clearly and faithfully as possible, much like a transparent windowpane providing a view of a scene. If the communication is done properly, the medium of communication is unobtrusive and contributes nothing of substance to the technology and information communicated through it. The ethical responsibilities involved in technical communication from this perspective are fairly clear and narrow: They are simply to relay faithfully information between transmitter and receiver. Implicitly, from this perspective, ethical responsibility concerning a technology is understood as attached primarily either to the technology itself or to the users of that technology and much less so to the communicators.

Increasingly over the last two decades, however, the ethical situation has become much less cut-and-dried. Technical communication, as we will see, has come to be understood as more complex, active, and creative than had been realized. Technical communication involves developing and creatively giving shape to information as it is presented effectively in various formats designed for the practical interests and needs of specific audiences. Seen from this fuller perspective, we stand in a different relationship to subject-matter experts, our clients and employers, our audiences, and society generally. To accomplish this work ethically, we as technical communicators need to understand the responsibilities that accompany this expanded role.

We have a responsibility to be familiar with and use the latest technologies of communication as well as with the technologies about which we communicate. We have a responsibility to understand the broad influence of technology and science throughout society. These responsibilities involve the uses to which our information will likely be put, the range of possible readings of our documents, and the consequences of our communications at all levels of society beyond the immediate audience. And we need to appreciate the impact of what we choose not to communicate. This is not to say that we have sole ethical responsibility in these matters, only that as our influence grows, so do our responsibilities. The purpose of this book is to help you—and our profession—better understand these fuller responsibilities.

WHY STUDY ETHICS?

We all continually face throughout our professional and personal lives the question, What is the right thing to do? Our answers of course depend on the particular dilemma we happen to be facing and its unique circumstances. But in general we each have a pretty good idea of what we will do in most situations. We know what we feel is "right," at least for us in most situations.

We might not, however, be able to articulate what makes a particular ethical judgment right for us, that is, the grounds for our ethical judgment. We also might have difficulty critically examining our own decision-making process and values. We might not be able to answer such questions as, Why did we decide this way? Are these reasons really good and adequate for rendering a responsible judgment, at least for me if not for others? Recalling Socrates' famous statement, "The unexamined life is not worth living," we might even wonder whether the ethically unexamined act is worth doing. Without a clear understanding of why we decided as we did, we might be unable to justify our decision. We might even be unable to persuade another person to support our decision or to render a similar judgment in a similar situation.

In certain situations our judgment might be less quick and less sure. What about complex situations in which a large number of important factors seem to be at work? For instance, an employer might want you to withhold or minimize potentially embarrassing information in a technical document. Do you rigidly fulfill your contractual obligations and do what the employer asks of you? Do you resist and vigorously defend a more honest and complete communication? Or do you refuse outright and threaten to blow the whistle and perhaps jeopardize your job and face the possibility of great negative consequences?

What about situations that are new to you but in which you nevertheless have to make an ethical decision? Industrial systems engineering might be totally outside your expertise, but you might nevertheless sense the need for a warning in a technical document you are working on. Faced with an imminent deadline, do you delay production to allay your qualms or just press on with the material as it was presented to you? What guidance can you turn to to make an ethical decision? Can you rely just on what others have done or said, such as in your organization's guidelines or in professional codes of ethics? Consider that Martin Luther King Jr. rejected even conformance to the law as the sole basis for ethical judgments because he distinguished fair, just laws with desirable consequences from unfair, unjust ones.

These plausible yet problematic ethical dilemmas indicate the need for us to seriously consider ethics for ourselves and our society. Throughout history, no serious thinker on the subject of ethics has ever concluded that a spontaneous, unconsidered impulse is ethically adequate. We need to examine the ethical bases of our decisions, decisions in general, and the decisions of others. We need to consider systematically what guidelines can help us to satisfy our ethical responsibilities to ourselves and to others. Although we might not uncover any universal, absolute dictates to guide ethical decisions such as a Ten Commandments of Ethics, we might at least come to better understand what we value and why. This self-awareness will lead us not so much to quick and sure judgments, perhaps, but at least to judgments that are more ethically satisfying because they have been carefully considered.

WHAT IS ETHICS?

What is the right thing to do? Why is it right? Why are some behaviors right and not others? Such questions deal with ethics, a field that has been the subject of both philosophical and practical investigation for many centuries. Ethics is a realm in which many people feel uncomfortable, however, partly because it has provoked conscientious speculation and contentious argumentation by many people yet has led to few if any concrete, absolute conclusions. Nothing in ethics corresponds, say, to Newtonian mechanics, which tells us exactly what will happen in a given case. If so many others have grappled with this subject without subduing it, one might well wonder, then how on earth can I?

The expectation that ethics should give clear and distinct answers is mistaken, of course. This expectation assumes ethics to be a subject like physics, which involves a body of knowledge about something other than ourselves. Ethics, though, is only about ourselves. As we will see, the large body of sometimes differing thought on ethics is only a reflection of the inherent nature of the subject and not of the limitations of our innate abilities, and so we should not be put off.

We should also recognize that we are bright, sensitive, earnest people who often face real ethical dilemmas and need to weigh our actions. As we will also see in this book, ethics is an area that has, paradoxically, no experts and yet many experts. Because the ethical burden falls on and must be borne by each individual person—even though many technical communications activities might be undertaken collectively in a social or corporate context—each of us is an ethical decision maker. For that reason, each of us is, in a sense, as expert as anyone else in judging what we ought to do. Regardless of all the ruminations of the great figures in the history of ethics, not one of them can take your place and face for you the unique particulars of your ethical circumstances. In this sense, each of us must become our own ethical expert and authority.

At the same time, the history of ethical studies makes it clear that, paradoxically, one should not make ethical decisions alone as a radical individual accountable only to oneself. Practically all ethical theorists indicate that each of us is responsible for weighing and choosing our options on the basis of some principle of responsibility that connects us all as human beings and as a society. In classical Greek times, this was the principle of wisdom attuned both to pleasing the gods and to living justly with our fellow human beings. In more recent times, it was the utilitarian principle seeking the greatest good for the greatest number, in which case social considerations outweigh self-interest. Nowadays it might be the principle of fundamental caring for those around us as unique individuals of equal importance to ourselves.

To enact thoughtfully our ethical responsibility as individuals, we need to understand what others have thought on the subject. We also need to understand what those who are affected by our decisions think and feel about ethical responsibility. We learn this by studying and communicating actively with those around us, much like a modern reenactment of the searching social debate exemplified long ago by the ethical paragon Socrates. The result of this studying, thinking, and discussing is that we better understand ourselves.

This book reviews the writings of several major ethical theorists to see how these views might apply to technical communication. It examines the ethical issues in a number of notable real cases involving technical communication.

It also explores and discusses the ethical issues in several hypothetical but realistic cases. Ultimately, the open, exploratory discussion of these cases, both real and hypothetical, is what this book is all about. Only by considering our own views on these cases and then discussing how others view them can we learn to recognize how complex, significant, and problematic ethical issues can be. Only then can we come to accept our responsibility for the difficult task of weighing such issues.

OUR EXPECTATIONS

Before embarking on our exploration of ethics, we should clarify our expectations of this exploration. To do this, we need to understand not only what ethics is, but also what ethics is not. Ethics is *not* many things, of course. To get a better grasp on ethics, it will be helpful to examine some of the assumptions that we might be bringing to our discussions. Being clear about both what to expect and what not to expect can help to prevent confusion and disappointment and perhaps increase our satisfaction with our discussion.

You might be able to see the importance of expectations if I mention my own concept of what ethics is not. Several years ago I participated in a faculty development seminar on applied and professional ethics at a large state university. Despite all that I had ever read about ethics, I confess that somehow I expected the seminar to make crystal clear for me what had been muddy. In thinking about ethics before this seminar, I had had a hard time deciding what I would actually do in ethically problematic situations. I expected that after participating in this seminar I would be able to make ethical judgments quickly and with great self-assurance. The reality was quite otherwise, of course, and that was the most important insight I gained from the seminar.

In retrospect I think that my deep educational background in the sciences, coupled with the whole weight of the scientific and technological mindset so characteristic of our contemporary culture, led me to expect that ethics could be treated somewhat like a science or technology. In the back of my mind, I expected to be presented with a shiny, new, simple yet powerful framework to displace the unclear, dusty, unending debates about ethics down through the ages. After all, a few simple equations from Newton enable us to plot with incredible precision the path of spacecraft orbiting Jupiter and sling-shotting around its moons. I expected to learn something like a computer program or algorithm for ethics. Plug in the initial conditions of the situation at hand, and the machinery would generate a clear and distinct—and ethical—decision for me.

I really wanted things to be clear and easy, a natural enough impulse but one that reveals my misunderstanding of the nature of ethics and moral reasoning. Only after discussing in this seminar various ethical systems, exploring concrete ethical dilemmas, and reading more deeply the works of ethical thinkers did I come to realize that ethics *is* problematic. Any inclination I might have to make things unproblematic and easy—though understandable enough—was erroneous and self-deluding. If we can understand a situation absolutely clearly and can reason out a single compelling decision on the matter, chances are we are operating not in the realm of the ethics at all but something else, such as logic or science.

As I read, studied, and discussed still further during that seminar, I came to realize that even the traditional emphasis on personal responsibility did not quite capture the whole picture of ethical deliberations. To the personal we must add the social circumstances as an essential component of ethical deliberations. The social circumstances allow for the input of sensible, responsible others. These inputs might be accepted or not by us, but the need to be open to different viewpoints is fundamental. Only when our deliberations are freely open to different voices can we legitimately say that we have arrived at truly ethical decisions. This, we will see, has been known as a precondition for ethics from the time of Plato and Aristotle to the twentieth century and thinkers such as Jürgen Habermas and Emmanuel Levinas. Only when local and global politics foster receptivity to a variety of attitudes and permit different voices to be heard can we make ethical decisions. For this reason ethics is inescapably linked to politics in the traditional civic sense and to open discourse, that is, to rhetoric.

ASSUMPTIONS

The word *ethics* will be used broadly but not loosely in this book. It can refer to the general field of study or to the theories of particular historical figures. It can also refer to the value system that anyone might hold, such as a personal ethic. It can refer, too, to the value assumptions of an abstract principle or movement, such as the ethics of science, as the system of values underlying it. In general, ethics has to do with values, with what we consider to be good, right, true, desirable, or commendable. It also has to do with the judgments we make in applying our values.

Values are one of the principal topics of this book. Values, we will see, underlie all communication, whether consciously or not. They often appear in the form of assumptions, whether tacit or explicit. I have listed the following principal assumptions about ethics as used in this book. Other books might adopt other assumptions, of course. Each of the assumptions listed here will be more fully explained later. They are offered here only as guideposts for understanding this text.

1. Ethics is problematic in several senses. It is about problems whose solutions are unclear at first sight. It is also about a body of thought that can only be suggestive, never dictating easy answers for us. Ethics in practice involves making a deliberate choice among alternatives in a real situation—a choice that is not always clear. Real ethical dilemmas are complex and usually quite difficult to judge. Do not expect easy answers; in fact, be wary

of them. Ethical deliberations are difficult, time-consuming work, but we should not for that reason shy away from them.

2. Paradoxically, ethics is both individual and social. It requires discussion, the talking out of issues among a number of people. But it also requires the assertion of individual responsibility, sometimes despite social pressures. Ethics is a matter of responsibility carried out at the personal level above all. When push comes to shove, individuals must assert their responsibilities, which can then be enacted singly or collectively.

3. Ethics is neither an entirely absolute nor an entirely relative matter. Few things, after all, can be known absolutely or their relevance to our immediate problems understood completely. On the other hand, it is usually not ethically responsible to say that any position on an issue is as valid as any other. Some courses of action are better in some senses and more right than others.

4. It would be irresponsible either to blindly accept or reject the authority of others in ethical matters. For that reason we will examine several ethical theories in order to learn from them. Some guidance is better than no guidance.

5. Due to the social, situation-specific nature of ethical judgments, no single ethical theory or approach will always be best for all situations. Instead we need to choose flexibly from the ones available to us while feeling free to add our own intuitions. For that reason this textbook emphasizes the free, open discussion of ethical dilemmas not for the sake of urging relativism but for the sake of developing a narrative, organic appraisal of your own that best fits both you and the situation.

PERSPECTIVES

The ethical perspectives we will explore are the standard ones that have received the most serious attention throughout the years. They have given us the terms we commonly use in judging human actions. They deal with virtue and character, duty and obligation, consequences in terms of costs and benefits, consideration of others, caring concern, and the quality of relationships. Generally these concepts will be used in the same way they are used in everyday conversation. When concepts are used in a specialized sense, they will be defined in context.

Time and space limit the number of ethical theories or perspectives we can consider here. I have chosen the ethical theories of Aristotle, Kant, utilitarianism, and recent feminist and care-based theories as covering the broadest range of perspectives receiving extensive recognition over the years. They have given us the concepts and terminology most people use in discussing ethics. Even though some of these perspectives—virtue, personal responsibility, duty, and obligation—are ancient, they are still meaningful. Ethics after all is as old as humankind, which faces many of the same issues now as in earlier times, though in newer contexts. Other ethical perspectives such as those of Levinas, Gert, and Confucius will also be discussed briefly.

The aim of this book is personal reflection and earnest intellectual discussion about ethics. It is not to urge one of the perspectives presented here over any others or to suggest that these are the only legitimate, meaningful approaches to ethics. Indeed, the multiplicity of ethical theories that has developed over the years can be so overwhelming that one hardly knows where to begin. Settling on a few of the more prominent ones at least provides a manageable starting point. At the same time, these few theories represent the conventional understanding of ethics. Although they might not be as intellectually challenging or sophisticated as other theories, they are readily accessible. They also utilize the language and concepts that are commonly used in everyday discourse about ethics, morality, and values and their impact on our lives. These theories are included only to precipitate reflection and social discussions about ethics.

Although they cover a broad span of time, they center on the European–American tradition including ancient Greece, Germany, Britain, and the United States. Confucian ethics, representing an older Chinese culture, is discussed briefly in one section but is not carried through to other sections in the case studies. Many other perspectives such as religious ethics or overtly political ethics or emerging perspectives such as environmental ethics are not discussed. Though the sampling presented here does not include the ethical perspectives of many other nations and cultural heritages, this is not meant to slight them. The focus of the book is ultimately how to deal with the ethical dilemmas faced by contemporary technical communicators, and so as a practical necessity we must address our discussions to the cultural tradition of the majority of our audience. As the world becomes increasingly technologized and evolves into a global community, multicultural sensitivity and accommodation are becoming increasingly important. Unfortunately, a fuller approach to multicultural ethics must await another book.

SCOPE

This book is limited in scope. No work other than the most unwieldy of tomes could hope to do full justice to all that ethics has to tell us as technical communicators. For this reason a narrow scope is necessary. I chose to focus on how ethics relates to technical communication in ways that are inapparent or nonobvious but no less real and powerful. These are the ways ethics lies in the background but still drives much of the substance of technical communication. This can be in the way technical information is obtained or will be used or how a quibble over a single word such as *cause* can have monumental ethical ramifications. It may be how unrealistic technical dreams can be represented as ethically questionable visions of possible realities and how technical feasibility can be mistaken for practical usefulness. It may also be how social forces can impinge on technical communications to totally distort the meaning of concrete technical terms such as *safety* and *flightworthiness*. These are usually situations in which the apparent technicality or scientific nature of the material heavily masks the ethical issues at work.

In limiting this book this way, other valid and useful approaches were necessarily left out due to the limitations of space. For the most part, the ethics involved in reporting data correctly, for example, are not included because this would deal with principles that apply not only to technical communication specifically but to a broad range of other fields as well. Matters such as copyright infringement are not addressed here because they relate more to legality than to ethics. Although the law is in some sense only a codified, institutionalized ethic, it nonetheless lacks the sense of a range of possible options from which to choose that is inherent to ethics. The federal government, furthermore, has extensive rules and regulations making ethics explicit and as concrete as possible. These enact a view of ethics limited to matters such as conflict of interest, financial disclosure, impartiality, and the appearance of impropriety. Other scholars are exploring interesting new ethical territory that has yet to be defined clearly such as the rhetorical ethics of networked communication.

ORGANIZATION

This chapter sets the stage for later chapters by outlining the topic of ethics as it will be explored in this book. Chapter 2 explains how ethics has been intimately tied to communication and rhetoric studies throughout history from classical Greece to the present-day United States. Every act of communication springs from and propagates an ethic, or value system. Chapter 3 focuses on the principal ethical theories of the European–American tradition, represented by Aristotle, Kant, utilitarianism, contemporary feminist ethics, and an ethic of care urged by some feminist thinkers and others. The ethical perspectives of Confucianism, Emmanuel Levinas, and Bernard Gert are also discussed briefly.

Chapters 4 through 7 apply the principal ethical theories to real cases of major ethical dilemmas of recent times involving technical communication in various ways. After the context, incident, and issues are explained, an ethical appraisal section applies the ethical theories to each real case. The real cases consist of the Nazi regime and the ethical implications of the origins and use of technical information; the shuttle *Challenger* disaster in the distortion of the meaning of technical information in technical reports; the American tobacco industry in the misrepresentation of the state of their knowledge, scientific interpretation of data, scientific principles, and the general understanding of the scientific community; and the Strategic Defense Initiative known as Star Wars in misrepresenting technical feasibility and the potential trustworthiness of software to operate this proposed massive defense system. The examples illustrate ethical issues of profound importance even though the ethical stakes might not be apparent at first glance. Although few technical communicators will ever face ethical dilemmas as momentous as these, the principles they illustrate are operative on a smaller scale in many everyday technical communication situations.

Chapter 8, on the other hand, presents six hypothetical but realistic cases illustrating ethical dilemmas that many technical communicators might plausibly face in their work. As with the real cases, an ethical appraisal applying the ethical theories follows each hypothetical case. All the chapters (after Chapter 1) include a section at the end on suggested topics for papers and class discussions and recommended Web sites.

TERMINOLOGY

In this book I will not usually distinguish between ethical and moral and will use the term *ethical* for both concepts. I will also usually not distinguish between ethics and values, with some exceptions. *Values* refers to the intentions or ends that guide an action, which need not involve the same sense of careful responsibility that is connoted by *ethics*. Ethics usually involves values, but values need not always involve ethics.

The terms *absolute* and *relative* are used frequently here as they are in other ethics textbooks. *Absolute* means definite, unchanging, and inflexible, applying to any and all situations in exactly the same way. "Written in stone" captures the feel of this term. We usually think of laws in a fairly absolute way. The law that requires us to stop at all red lights means just what it says, regardless of who is driving, what time of day it is, whether the driver is having a good day or a bad day, or what kind of car is involved. (The enforcement of the law might not be quite so rigid, however.) *Relative* means changing in relation to circumstances. Speed limits often include adjustments for local conditions such as snow, fog, or darkness. Even those states with no speed limit on some roads still qualify that limitlessness with terms such as *prudent* or *due caution*. A relative ethics would be an ethical system that depends on the particular circumstances, such as the particular person involved, a person's motives or intentions, or the severity of the consequences.

Sometimes "relative" in ethics can be carried to an extreme. This happens when the ethical guidelines become so flexible and indefinite that it seems that anything goes and one is free to define criteria to suit oneself (and *not* to suit anyone else). It becomes, in effect, no ethics at all. Few serious thinkers take ethical relativism that far, although they might appear that way to critics. Other terms will be defined as they appear.

CHAPTER 2

SURVEY OF ETHICS IN COMMUNICATION AND RHETORIC

Twentieth-century ethical issues in technical communication are best understood by viewing them through the lens of history. Oftentimes viewing something from a distance—in this case the distance of time—allows us to see the overall forest when we would otherwise be too close to the trees. This historical review shows how ethics has been understood as inseparably linked to rhetoric, what we nowadays call communication.

What does Socrates say about full communications open to criticism and about the ethical responsibilities of the communicator? Would Plato condone technical writing that does not scrupulously point out hazards but only leaves them for the reader to infer? Would Kant approve of impersonal memos that disregard the persons affected by these technical communications? What does Aristotle say to us about how to write honestly about technical matters? What does Foucault reveal about the connections among power, values, and knowledge?

As we will see, historical philosophers can help us to grasp the ethical dimensions of technical communication. What we write does have consequences, and we must accept responsibility for our words. The same is true for our images, sounds, Web sites, or any other elements of communication. Thus our communications must involve thoughtful decisions about intentions, consequences, and values. Our social engagement with those around us can help us to make these decisions more intelligently.

In this chapter we see how the relationship between ethics and rhetoric as communication has been viewed in classical times. We also see, perhaps surprisingly, how these classical views resonate with the latest views on the matter. This review is important because in classical times many of the fundamental concepts and issues that have long characterized ethics as a field of study were first raised and addressed. They were studied because they are so fundamentally important and they occur so frequently. And they continue to be important and frequently encountered.

Seeing these topics at arm's length lets us understand them more clearly than we might if we were to consider them only within our own times. Just as travel to other countries allows us to better appreciate and understand our own society, so can the study of past thinkers help us to develop our own thinking. Among the topics we will explore are the nature of right and wrong; the sources of our standards for ethical judgment, whether human, natural, metaphysical, or divine; individual versus social interests; the significance of intent; reasoned deliberateness versus emotional impulsiveness; and the role of rhetoric and persuasion in the social negotiation of value judgments. We also address questions such as, What is the relative importance of intentions and consequences? Are usefulness and effectiveness good standards for making ethical judgments? What is the best course of action in a given situation? Why? And for whom? These fundamental topics continue to be important for our contemporary ethical circumstances, though in updated form.

In this book we do not use the terms *rhetoric* or *persuasion* in the pejorative sense they sometimes have as deception, manipulation, or control. Rather, we use these terms in their best sense. Rhetoric means the use of reasoned arguments based on socially accepted values and presented to inform and persuade in order to accomplish some socially desirable action such as a policy decision. Persuasion means the willing, informed collective agreement of a critically thinking audience.

How is ethics linked to rhetoric? As we will see, most rhetorical thinkers and philosophers believe ethics and rhetoric go hand-in-hand: You cannot have one without the other. Rhetoric is always the urging of a point of view, or judgment, or course of action. It always presents good reasons for its urgings, and these reasons amount to a system of values, whether explicit or implicit. Ethics, on the other hand, is the systematic study of the problematic. It aims at clarifying an unclear situation. But even our study of ethical history can carry a note of advocacy and persuasion by assuming that we should perhaps accept the insights of the ethical thinkers. After all, that is what systematic study always aims at—a general awareness that applies to other people and situations. Because human communication is always about values and always rhetorical at some level, we should strive to be conscious of the ethical values communicated through the rhetoric of our discourse.

We will see in later chapters how the insights of both historical and modern thinkers on ethics and rhetoric are played out in many modern technical communication situations. Regarding Plato, for example, we will see the central importance of pursuing what is true versus trying to avoid the truth and the important role of society in protecting the health of its citizens. We will see that Nazi pseudoscientific research was simply not right, as was determined by an international tribunal of judges. Regarding Aristotle, we will see examples of virtues such as honesty and a concern for the public welfare displayed by Roger Boisjoly in the *Challenger* disaster, by William LeMessurier in the technical documents of a skyscraper he designed, and by David Parnas in opposing unrealistic claims about the proposed software for the Star Wars missile defense system. The ancient sophists, who claimed that every topic has several sides that can be legitimately argued about and who were famous for standing conventional thinking on its head, are represented, too. The tobacco industry has argued that there is a controversy about the health effects of smoking when in fact there is no controversy among practically all respected scientists without ties to the tobacco industry. We will also examine testimony in the U.S. Congress in a debate that deliberately brings two opposing views into confrontation in order to form a judgment.

Turning to modern times, we will see that many contemporary thinkers consider science and technology to be value systems of sorts even though they appear not to be. They are value systems because they have not only criteria for establishing knowledge but also play important roles in society in settling disputes and because they are pursued for their own sakes. The Star Wars program is a good illustration of the interconnection between facts and opinions, in this case technology and political goals. The program, for example, was based to a large extent on claims about what technology could do if that technology were invented or discovered. Its technical claims, to a large degree were actually dreams and hopes founded on the assumption that technology will give us what we need if only we pursue (and fund) it vigorously enough.

LIMITATIONS OF HISTORY

The fundamental issues raised in the classical period continue to be valid and operative in our own times, but we should understand that historical views are relevant to us only in broad terms. Although the free, open, and conscientious discussion of ethical issues has always been vitally important in making responsible judgments, we need to understand that these historical discussions never have resolved any of these issues once and for all. Because we cannot mechanically adopt past decisions to present situations, we cannot relieve ourselves of the need to weigh ethical issues by simply deferring to ancient authorities. We cannot, that is, expect our historical study of ethical theory to do away with questions, problems, and dilemmas. Even though history and theory can both clarify issues and problematize them, it cannot make decisions for us. What we learn from past ethical judgments is only a fuller understanding of the complexity and contingency of real ethical dilemmas and a fuller understanding of the vital importance of continual, open communication about ethical matters—but this is a great deal.

As we will see, the very nature of ethical issues is to be problematic (explained further in Dombrowski's "Technologize"). For that reason no ethical approach should ever be expected to settle things once and for all. In any particular ethical situation, even though we might firmly decide on a particular course of action to take, our particular decisions at that time can by no means relieve us of the responsibility to discuss continually ethical issues in any and all other situations. Therefore our study of the history of ethics cannot give us universal solutions for our ethical dilemmas. But history can show us how others have conscientiously dealt with particular issues and also show us some of the complex dimensions of ethical issues that might otherwise escape our notice. It can also provide clear examples of an attitude of willingness to grapple with difficult ethical issues.

ETHICS AND RHETORIC LINKED

In this section we trace the history of the relationship between ethics and rhetoric as major thinkers have understood it in different eras. This review of history is important because this book is about the interrelation between ethics and technical communication and because communication necessarily entails rhetoric. Thus this book is only a recent addition to a long and well-respected tradition linking ethics and rhetoric.

We will see that how technical communication is understood as rhetorical or not is a fundamental question about our field. We will also see that this question is closely linked to how technical communication is understood as involved with ethics. In general those who feel that technical communication is not at all rhetorical usually also feel that technical communication has not a great deal to do with ethics because it involves only relaying factual information. Technical communication for them is ethical to the extent that it relays information correctly, accurately, and clearly. On the other hand, those who define rhetoric and ethics more broadly feel that technical communication is always rhetorical and always has to do with ethics and values. This book takes the latter position.

Rhetoric refers to all manner of persuasion, argumentation, and negotiation in communication, whether oral, written, visual, or electronic. It deals with how and why we use language, including the ideas, thoughts, opinions, and values embodied in language. It also involves the things, processes, and goals that language refers to in technical communications. As we will see, many modern thinkers feel that technology is a value system or at least a value choice. It prefers to deal with the objective rather than the subjective. It also prefers to communicate about technology, its artifacts, and its information rather than about feelings, aesthetic tastes, or political preferences.

Classical Greece

The roots of the systematic study of communication as human interaction, known as rhetoric, lie in classical Greece. In this section, we will examine the thinking of Plato, Socrates, Aristotle, and the sophists on the relationships between ethics and rhetoric as communication. In a nutshell, for Plato ethical values come before any communication; for Aristotle the communication between competing sides on a controversial matter reveals the proper values and the right course of action; and for the sophists the communication act can alter our ethical values because there is no absolute basis for ethics.

Even though at times the connection between these ideas and modern technical communication might not be immediately clear, keep in mind that our purpose is to develop fundamental concepts of lasting importance. We will see, for example, that the question of whether ethics is connected to technical knowledge at all has been debated for a very long time but has yielded no easy answers. The connection at times has been perceived as very tight but at other times as very loose. The end of the discussion of each era presents specific conclusions relating to technical communication.

Plato and Socrates.   Plato is commonly considered the founder of philosophy. He held that philosophy is a matter of discovering and pursuing truth, goodness, and rightness. Ethics is the branch of philosophy concerned with determining right conduct. Plato's philosophy is theological, and for many of us his focus on god might seem at first irrelevant to modern technical communication. By the end of our review of Plato, however, the connection to technical communication will probably seem clearer and more important than at first.

We humans are unique, Plata believed, because of our knowledge of and participation in the divine through our soul. As humans we should follow where we are led by our soul, which, as a piece of the immortal godhead within us, innately seeks to be reunited with god. (Plato's god is not the same as the Judeo–Christian God, of course.) This is where our higher, better nature leads us, so this is where we should follow. We should strive to please god, and in so doing we will necessarily please ourselves because that is what our souls really want.

Because ethics is a matter of pleasing god, who can only be immortal and unchangeable, Plato says that ethics must also be unchangeable and not subject to contingencies. Thus Plato's ethics is absolutist. (Absolutism means that things are only one, definite way that never changes.) There is only one right way of acting in any given situation, which is a function not of the situation itself but only of fundamental, absolute laws. Furthermore, determining the right way of acting is not a matter of wholly human judgment, such as taking a majority vote in a mass meeting, but of conscientiously trying to understand the will of god on the issue.

Plato's absolutist ethics is largely a reflection of his education, personal experience, and heartfelt feelings, in a powerful way. Plato was a pupil of Socrates, whom Plato revered. Socrates was a social critic and teacher who insisted that "the unexamined life is not worth living." From his perspective, our innate burden is to continually reflect on and critically examine all aspects of our lives because only through this conscientious reflection can we determine what is right for us to do.

For Socrates ethics and right living was a very serious and never-ending activity. Our ethical burden is carried out through examination not only within ourselves but also socially because of our responsibility to god through others. Our responsibilities to others are primarily motivated by their likeness to god and therefore to us ourselves. For Socrates, the social interchange does not define what is good; rather, the social dialogue makes possible divine enlightenment for all concerned. For him, this ethical burden was very serious. He continued to conduct his social criticism and was eventually put on trial for it and condemned to death.

For our purposes, Socrates is important for three reasons. First, he insisted on doing the right thing regardless of the consequences. He took his ethical responsibilities as more important than life itself (actually, for him it was life). Second, ethics is a matter of pleasing god. Though he was a practical man who participated in both military and political service to Athens, Socrates had a spiritual side that was vitally important to him. He insisted on following his conscience as it was led by the will of god, regardless of what any of the fallible people around him happened to think he should do. In this sense, ethics was an absolute matter that should not be influenced by circumstances or personal opinions. In this way, as we will see, Socrates fundamentally differs from some twentieth-century ethical positions. Third, ethical behavior requires active social involvement, whether as teacher, civic leader, or social critic. Thus Socrates was continually engaging the most important and powerful people of his time in intense discussions about the right thing to do. More important, ethical conduct for him *required* communication because we relate to god through our relations with fellow human beings, and we relate to others for practical effect through our communications.

The totally unjust persecution and execution of Socrates greatly moved his prime pupil, Plato. As a response to this gross injustice, Plato developed his own institute of higher education, called the Academy, the first and most famous of the ancient schools of higher education. Plato's ethical theory, not surprisingly, was entirely religious and absolutist. So, although Plato is often dismissed as an unrealistic idealist, we should remember that his ethics and rhetoric were his response to personally meaningful real events.

His ethics was authoritarian because only the brightest, most sensitive, and most conscientious people could have a clear perception of the will of god so as to guide human conduct. These people should lead while the rest follow. The role of rhetoric in Plato's view was to relay one's awareness of the divine, transmitting and transferring this absolute knowledge to those who lacked it. This makes communication largely a one-way affair, from the enlightened to those in need of enlightenment. In this way Plato differs somewhat from Socrates. For our purposes, we should see that this is not a rhetoric of social negotiation and critical analysis but a rhetoric of privileged authority.

Despite this sense of superiority, known as elitism, running through Plato's thinking, his most important contribution to the history of rhetoric is his insistence on the ethical goodness of the communicator. Good communication first and foremost must be ethical and deal only with what is true, good, and right. This can occur only when the communicators fully understand both the topic being communicated and the nature of goodness and practice it in their own lives. As George Kennedy, the famous scholar of classical rhetoric, points out about Plato's theory of rhetoric, "Its main strength is the insistence on knowledge as the true basis of valid communication" (51). Kennedy also contrasts Plato to other rhetoricians such as Isocrates and the more responsible sophists. Because these, he says, were willing to conceive of rhetorical practice as separate from the qualities of the communicator, they appeared to tolerate or even condone evil. Plato, on the other hand, insisted on the "integration of the intellectual, moral, and rhetorical qualities of the orator into the whole man" (52).

To illustrate the vital importance of moral goodness in Plato's theory of rhetoric, consider one of his most complex dialogues, *Phaedrus*. Plato, through the character Socrates, explains the importance of knowledge of ethics and moral goodness. He also explains the equal importance of a genuinely loving attitude in the relationship between the communicator and the audience, which in turn reflects a love of goodness and truth. Only then can true communication take place, which will necessarily be focused on truth above all for the good of the audience. Plato's view of communication in the dialogue *Phaedrus* insists that we should learn what is right before we begin communicating to ensure that we say only what is right.

In technical communication studies we are not used to speaking of love and spirituality. Nevertheless, the lesson of Plato's compelling analysis of the vital importance of attitudes and values in the caring relationship between communicator and audience remains as valid now as it was then. This attitude of Plato's, we will later see, is not very different from that of many later ethical thinkers. Kant, for instance, emphasizes duty and respect for others, regardless of personal consequences, just as Socrates did. Some feminist ethical thinkers emphasize a genuinely caring relationship as the cornerstone of ethics. And like modern technical communication teachers, Plato and Socrates emphasized that the purpose of our discourse must always be the communication of the full truth as convincingly as possible. They also emphasized that discourse must respond to the specific, unique, immediate communication situation. For that reason, what we do in one situation does not necessarily apply for other situations, which leaves us with the burden to continually and consciously consider rhetoric and ethics anew.

The Platonic connection between ethics and rhetoric is illustrated in this book in several examples from the twentieth century. In the controversy about the health effects of smoking, opponents of smoking insist that it is true that smoking causes cancer and other disorders and that the government is right to take steps to protect the public good. Some of the Nazi documents we examine report on pseudomedical experiments on death camp prisoners, which absolutely were not right, as was determined by an international military tribunal of some of the world's most respected judges.

Aristotle. Aristotle, Plato's student, was more practical and less metaphysical than Plato, though he still maintained that ethics ultimately stems from the divinely ordained nature of things. The right course of action in a given situation, however, often cannot be *known* clearly. Therefore we often have to rely on fallible human judgment to determine the right course of action. This is where communication and rhetoric come into play. For this reason, according to Aristotle, "Rhetoric is an offshoot of dialectic [logic] and also of ethical studies" (25).

Aristotle articulated what has come to be the most famous classical definition of this art: "Rhetoric may be defined as the faculty of observing in any given case the available means of persuasion" (24). In classical times the practical art of rhetoric was usually played out either in the law courts or the legislative assemblies, roughly equivalent to our courts and Congress. Typically these rhetorical activities involved trying to persuade an audience to render a judgment favorable to one's side of the case.

Though Aristotle believed that ethics ultimately stemmed from the essential order of things, in practical affairs the ethical course of action had to be determined in a debate. Aristotle believed that, given a fair, even-handed contest between people arguing for competing courses of action, the innate goodness of one of the courses would naturally assert itself, much like cream rising to the surface of a mixed suspension of natural milk. "Rhetoric is useful because things that are true and things that are just have a natural tendency to prevail over their opposites" (22).

We should recognize, however, an important qualification when discussing Aristotle on ethics and rhetoric. For Aristotle there were two categories of knowledge: certain knowledge and uncertain knowledge. An example of certain knowledge is the Pythagorean theorem; an example of uncertain knowledge is whether Sparta will attack soon. Rhetoric for Aristotle deals only with what we do not know with certainty such as opinions, beliefs, and likelihood. Science, on the other hand, deals with knowledge that is true and certain. No one debates whether an apple falls downward or upward.

Aristotle's views are a mixed aid for us, then. On the one hand, Aristotle explains the link between ethics and rhetoric. On the other hand, he muddies the waters by distancing rhetoric from science and presumably from technology. If we were to accept Aristotle's position, then the certain knowledge that comes from empirical science and technology would be entirely separate from the realm of ethics and rhetoric. As we will later see regarding the twentieth century, however, contemporary thinkers find very good reasons for rejecting this supposed separation of rhetoric and ethics from science and technology. The modern view is that science and technology are systems of rhetorically negotiated knowledge, and they are systems of values.

Aristotle viewed ethics basically as virtue, whether personal or corporate. Ethics as virtue will be illustrated in some of the technical documents we will examine. The virtue of honesty seems clearly to be lacking in the arguments by the tobacco industry in defense of smoking, as does the virtue of civic responsibility for the welfare of the nation's citizens. Roger Boisjoly, an experienced rocket engineer, argued vigorously against the planned launch of the shuttle *Challenger* and lost his job. William J. LeMessurier, a highly respected architect, conscientiously brought his own errors in the design of a newly constructed skyscraper to the attention of the authorities, the same authorities who had approved his plans in the first place. David Parnas, often called the father of software engineering, opposed many of his colleagues in stating that the proposed software for the Star Wars missile defense system could never be made trustworthy.

The Sophists.   The sophists represent a position rather different from Plato's on ethics and its relationship to rhetoric. The sophists were not an actual school such as Plato's Academy or Aristotle's Lyceum or even a cohesive movement but a loose category of emerging freethinkers and teachers. There is even serious disagreement about who was a sophist and who was not. Little is known of them in their own words because only a very few fragmentary pieces of primary materials survive.

What we do know about them comes instead from others writing about them. This secondary material is almost entirely critical or even hostile toward the sophists. Chief among these critics was Plato himself, who attacked them strenuously and often. One wonders whether this gives us a valid representation of what they actually said, did, or stood for. In the last century or so, a number of scholars have challenged the traditional view that leads back to the time of the sophists themselves. They propose that the sophists have been severely and inappropriately maligned, and the predominance of Plato and Aristotle rather than the sophists in our intellectual heritage is the result of suppression, active exclusion, and malicious disempowerment rather than of any innate shortcoming of the sophists' position. Susan Jarratt, for instance, urges the "refiguring" of the sophists so that we can more truthfully understand their philosophy, principles, and motives. Feminists in particular but not only they, Jarratt says, can learn from the sophists the techniques for opposing and revamping an unfair social order based on the fundamental recognition of the power of language to shape values and thoughts.

The animosity between the sophists and Plato in particular is legendary and had much to do with ethics and rhetoric. Plato, through the character Socrates in his dialogues, repeatedly attacked the sophists as representing the worst sort of values and communication. Protagoras, one of the chief sophists, and other sophists, on the other hand, claimed to have the power through their rhetorical craft to make the weaker case appear the stronger, or the worse case seem the better. Obviously their rhetorical discourse stemmed not from absolute truth and perfect enlightenment but from some other basis. Socrates and Plato, on the other hand, felt that values are absolute because they come from god and so must be learned and then applied to communication. The sophists generally held the opposite view, that there are no absolutes and that communication is immensely powerful precisely because it shapes minds, hearts, values, and decisions.

The sophists insisted on the fundamental indeterminacy of everything we take as knowledge whether it be scientific knowledge or political beliefs. This is because language, according to this position, does not refer to or represent anything that exists before and separate from our language use. Instead, language use, as a rhetorically negotiated social construct, itself defines what is to be taken as things and in so doing constitutes meaning itself. Our words refer not to things but to other words, to our entire language system, and to the constellation of meanings in that language.

The sophists typically held what we now call a cultural relativist position on ethical values. This asserts that values are relative because they depend on circumstances. In one culture, for example, a gesture might be good while bad in another. The sophists asserted that because cultures differ as to what is true, right, and good, we must take any and all of these values as equally valid. We must therefore reject the idea that only one is absolutely true, right, and good. The representation of the sophists by W. K. C. Guthrie, for instance, is fairly traditional (though this view has been challenged in recent years):

> Appearances [for the sophists] are constantly shifting, from one moment to the next and between one individual and another, and they themselves constitute the only reality. In morals this leads to a "situational ethics," an emphasis on the immediately practical and a distrust of general and permanent rules and principles . . . [By contrast] . . . in Plato's ideal theory, . . . such concepts as justice and beauty, as well as identity and equality and many others have an existence apart from the human mind, as independent and unvarying standards to which human perceptions and human actions can and must be referred (4).

In their cultural relativism the sophists were very much like our modern social constructionists. Social constructionism is the point of view that all knowledge is only a construct deriving from its social context. It is therefore susceptible to alteration because it is only a reflection of the prevailing social interests. In relation to science and technology in particular, it asserts that such knowledge is not the collection of absolute objective facts that it may appear to be but only a reflected sort of subjectivity. Peter Berger and Thomas Luckmann, in *The Social Construction of Reality*, are among the most important theorists of social constructionism. These sociologists argue that what we call "reality" must be socially constructed even though it is commonly assumed *not* to be socially constructed. Because people in different cultures inhabit very different "realities," all of which are assumed to be valid and workable, there really is no single, absolute, true reality, only a diverse collection of similarly valid realities. These various realities reflect the societies from which they come and do not reflect an absolute existence. Neil Everden, in *The Social Creation of Nature*, takes a similar view, explaining that even what we call "nature" varies from culture to culture and so cannot be the absolute reality we usually assume it to be. Richard

Rorty, a contemporary American pragmatist philosopher, is among the most renowned philosophers taking this position regarding all knowledge. For these scholars as for the sophists, language is everything, even to the extent of molding our ethics and changing what is valued depending on the circumstances.

Plato held that there are absolute standards of rightness because they originate in god and that these could be known if only we try. To be sure, only a few select people were able and willing to make the effort to arrive at such knowledge, but it could, in the end, be known. The sophists, on the other hand, held a much more skeptical view. Protagoras, for example, said pointedly that he did not know whether the gods exist at all, and he presumed to offer good reasons for thinking so. Gorgias, another leading sophist, held a highly skeptical position, contending that he did not know whether anything at all really exists in any absolute sense; or, if it does exist, whether it can be known by us; or, if it can be known, whether that knowledge can be communicated to others. All we have, according to the sophists, is communication among ourselves about what to think, believe, and do, and we should make the most of this. "The idea of law as no more than an agreement, instituted by men and alterable by consent, is . . . basic to the humanism of the Greek Sophists . . . ," Guthrie explains (6). From the sophists' perspective, it is the ethical responsibility of all of us to learn to speak as effectively and persuasively as possible because our ethics is created by our communications.

The traditional view critical of the sophists is not the only respectable view, however. Edward Schiappa, another prominent scholar on the sophists, points out that the sophists are often misrepresented, especially in their supposed opposition to Plato and Aristotle. He points out that Protagoras actually advocated a civically responsible use of rhetoric in relation to ethics, arguing that rhetoric could act very much like medicine as a corrective to mistaken social values and opinions. "For Protagoras, *logos* [i.e., reasoned argument] was the means through which citizens deliberated and came to collective judgments. Protagoras contributed to the theoretical defense of consensual decision-making, and he may have been the first to provide rules to facilitate the orderly conduct of debate and discussion" (*Protagoras and Logos*, 199).

Although Schiappa has high regard for individual sophists, he wonders about the validity of speaking of the sophists as a unitary entity and of using them as representatives of disempowerment and as authorities on democratic social criticism. Most were very well off financially as a result of their teaching, and several were attacked for their antidemocratic notions, for example. "The attributes one finds common to all or most of the standard list of sophists are also common to many other thinkers of the fifth century [BCE] . . . When one finds an authentically unique contribution, such as Gorgias' account of the power of *logos*, the very uniqueness that makes the contribution noteworthy makes the label 'sophistic' an over-generalization" ("Oasis or Mirage," 10). Instead Schiappa, who is generally sympathetic with the critical notions attributed to the sophists by recent rehabilitators of them, argues that these notions are worthy of standing in their own right rather than on questionable historical grounds.

Another notable voice on the complex history of representations of the sophists is G. B. Kerferd, whose book on the sophistic movement is a standard. Many notions attributed to the sophists, he points out, are highly problematic—not entirely wrong but often overstated. Nevertheless, Kerferd acknowledges that many of the famous philosophical notions attributed to Plato and Aristotle derived from intellectual investigations begun first by various sophists. In addition, many fundamental philosophical issues of lasting—even current—interest were first brought to prominent attention by the sophists. These include "the theory of knowledge, . . . the nature of truth, . . . and the relation between language, thought and reality" (2). One of these issues was whether the basis for social order should be *nomos*, the traditional laws and customs created by mankind, or *physis*, nature and its principles. This interplay between human society and natural laws (such as from technology and science) is much the same intellectual territory in which technical communication operates today. Underlying all these activities is the fundamental assumption that language is powerful, that it is tied to no absolutes, and that it shapes the ways we think, believe, act, and know. They also initiated spirited public debates on these topics to which Plato and others reacted. It is fair to say that of all the activities that can be attributed to the sophists, perhaps the most basic and yet the most important was debate itself—active, continual, provocative, and disinterested though rarely uninteresting.

Some things can definitely be attributed to particular sophists. Gorgias was famous for his ability to persuade through communication, even to persuade an audience against their deeply held opinions and value judgments. Gorgias's most famous speech is his *Encomium of Helen*, which served to showcase his persuasive power to shape values. Helen of Troy was the wife of king Menelaus of Sparta. She left him for general Paris of Troy. Her leaving and the Greek efforts to reclaim her brought about the famous Trojan war. Thus the poet Homer says of Helen that hers was "the face that launched a thousand ships." Helen was traditionally reviled as an adulteress and a traitor. In his *Encomium of Helen*, however, Gorgias set out to persuade his audience that Helen was a helpless victim of emotions and of Paris's greater strength and persuasive power. Thus, through his speech, Gorgias urged the audience to appraise Helen in a much different light and to judge her ethically in a different way. The speech therefore demonstrates the power of communication in general, and of Gorgias' own abilities in particular, to shape the values that we apply in our judgments. In effect, the communication act shapes the values. This is a much more active and flexible understanding of ethical values in relation to communication than Plato's view, in which communication is only a passive vehicle for the transmission of preexisting values. For Gorgias the values stem from the communication, while for Protagoras any supposed metaphysical truths could never be known in the first place.

An additional significant difference between the two camps, Plato and the sophists, concerns ethics itself. Plato (and Socrates) held that teachers have a responsibility to instill both ethical values and rhetorical arts in their pupils. The sophists held the opposite view, namely that rhetoric is only a skill, a collection of techniques—in effect a technology—that could be readily acquired by anyone. The learning of these rhetorical skills was in no way connected with the learning of ethical values.

The traditional view of the sophists concerning the connection between ethics and rhetoric is represented in this book in several documents. The tobacco industry's argument that a massive body of scientific evidence linking smoking to cancer and other disorders did not actually prove that smoking causes cancer and illness is very much like the sophists' making the greater point seem the lesser. In this case, they argued that a cause is not a cause. Their public denial of what they privately knew and everyone else had already accepted as known is also sophistic in assuming that there are two legitimate sides to any matter. The Nazi documents on anti-Semitic racism also are sophistic (at least in the traditional meaning of the term) in fabricating pseudoscientific support for their policy of annihilating the Jews, making what is not science appear to be science. In addition, their policy of aggressively propagandizing and of cutting off conventional public discourse had the effect of making often-repeated claims appear to be established truths. From Plato's perspective, this would be a sophistic ploy.

The contemporary "rehabilitative" view of the sophists would perceive these documents in a different light. Regarding the tobacco document, they would focus on the fundamental indeterminacy of knowledge and the need for earnest, open social debate to establish what will be assumed true on a given subject. From this view, the controversy referred to and instigated by the tobacco industry was a valid position and a reasonable rhetorical strategy reflecting a sincere perception. There simply is no single true or right position on the matter; any issue is *assumed* from the very beginning to have two sides.

To summarize, for Plato ethics and rhetoric are closely tied, but ethics comes first, and the only purpose of rhetoric is to serve ethics. For the sophists, ethics and rhetoric are also closely tied, but rhetoric comes first because it allows the negotiation and persuasion that defines social values.

Recent Times

Let us jump ahead about twenty centuries to more recent times, after briefly reviewing the intervening years. Between classical Greece and renaissance Europe, Aristotle's view of ethics and communication predominated. Plato's view, to a lesser extent, continued to be highly influential, too, but in the form of religious ethics, specifically the views of the Christian Church. Basically, Plato's metaphysical and absolutist views were preserved in the teachings of Saint Augustine, whereas Aristotle's pragmatic and argument-centered views were preserved in the later teachings of Saint Thomas Aquinas. The sophistic view was practically nonexistent except to demonstrate by contrast the assumed preferability of the Platonic and Aristotelian views. In these years ethics *was* Plato and Aristotle.

And ethics was a religious matter. The authority of the church was paramount. During the middle period—covering roughly the end of the classical period and the beginning of the renaissance—rhetoric was largely divorced from ethics. Ethics had to do with truths, whereas rhetoric had to do with false, insubstantial appearances. Although a form of rhetoric specific to religious preaching, homiletics, was practiced in the church, this was not really considered rhetoric. Rhetoric was considered to be something added on to the real substance of discourse. It amounted to empty ornamentation at best or deception and manipulation at worst. Indeed, for the most part rhetoric was considered the opposite of ethics. This period, then, marked the low point of opinions about rhetoric and about the connection between ethics and rhetoric.

In the seventeenth and eighteenth centuries, following the renaissance revival of classical Greek thought, a renewed recognition of the powers of reasoning developed throughout Europe. This view highlighted our abilities to reason about anything and everything, including ethical matters. This new recognition was especially important because it undermined the prevailing view about the source of ethics, namely, authority figures. These figures could be God, the church fathers, or even the great classical thinkers Aristotle and Plato (as interpreted suitably by others). At about the same time as the rise of rational inquiry, modern science was also developing. Insisting on the primacy of the here and now, it compounded the new rationalistic challenges to received authority, including ethics.

Hegel. In the nineteenth century G. W. F. Hegel is important for our study of ethics for his strident opposition to the prevailing view that all knowledge, especially about values, was absolute and had to be arrived at by revelation to accepted authorities. This knowledge was then handed down to the common people. Hegel was a radical skeptic who sought an objective understanding of the world as we experience it. He felt that values were arrived at socially, often by forces that were working to preserve only their own self-interest in what amounted to power plays. Values therefore do not derive from the absolute but rather from social forces. More important, Hegel asserted the powerful role of communication itself in the overall social process whereby values are proposed, endorsed, and propagated. Not surprisingly given this view that values are contingent, changeable, and actively shaped through communication, Hegel is famous as a modern "rehabilitator" of the sophists, rescuing them from traditional ridicule. He elevated them to a position on a par with Plato and Aristotle as among the greatest thinkers of their times. From Hegel's viewpoint, because the ethics and rhetoric of Plato and Aristotle are largely responses to the sophists, clearly the sophists' teachings were taken seriously even then. Therefore the sophists were implicitly acknowledged by Plato and Aristotle themselves as important thinkers.

Perelman. In the twentieth century Chaim Perelman explored the social contingency of values and the fundamental role of rhetoric in establishing values. Perelman was motivated by his judicial experience to examine the foundations of the values that underlie judicial judgments. He was surprised to find that there exists no absolute basis or authority for values, whether legal or ethical. Instead, he found, in effect, that nothing absolute lies behind or beyond rhetorical language use to serve as an authority or foundation for our values. Rather than language expressing or representing principles that are of a transcendent or metaphysical nature, our language, Perelman came to believe, *is* our values. Thus, when we communicate or participate in any rhetorical act, we are expressing values, of course, but also asserting them, offering them for consideration, and reinforcing them. Our values are already constituted in language. When we seek to persuade an audience, we are actually offering for assent our statements about values. Persuasion, then, amounts to a process of social negotiation, ratification, and propagation.

Even purely logical reasoning, according to Perelman, is rhetorical and fundamentally concerned with values. It is rhetorical in aiming to persuade an audience to assent to its conclusions, but this can occur only to the extent that the audience values carefully reasoned argument in the first place. Likewise, technical and scientific reasoning would be persuasive only if the audience already valued rationality and empirical demonstration. Another audience could value other, nonrationalistic ways of arguing and persuading and so would be impervious to the force of logic. Perelman explains that what usually occurs in persuasion is that the real audience harbors an ideal image of themselves as they would like to be. The real audience is persuaded to the extent that their ideal values are activated by the discourse.

Burke. Kenneth Burke is another important rhetorical theorist of the twentieth century. Burke insisted on language use guided by carefully weighed judgment. Nonetheless, it is true that Burke is considered one of the most important of modern social constructionists because he showed the fundamentally social nature of all language use and the power of language to negotiate knowledge and value judgments through rhetoric. The sophists, similarly, considered language to be an act of social construction. They thought, too, that our society and all its values are social constructs negotiated through language. Therefore values are worked out socially and rhetorically rather than received from on high. They also held that rhetoric worked to shape value judgments. Burke is a complicated thinker. He was not an idealist exactly, though he valued transcendence highly. Neither was he a relativist, though he acknowledged the role of social negotiation in shaping values. It is fair to say that he insisted on the social ratification of values but not that values originate solely from society.

Burke is also known as a symbolic interactionist because he understood language to be a symbol system. Language and its symbols refer not so much to things in the external, nonhuman environment but primarily only to language itself, reflexively. Our words refer not to things but to other words and to concepts embodied in words. All we have is our words, our language use among ourselves—but that is a great deal.

Weaver.   Others in the twentieth century were also concerned about the connection between ethics and communication, though in a rather antisophistic way. One of the most famous of these was Richard Weaver. Weaver is known for his essays on the rhetoric of Plato and for his modern explanation of the role of ethics in communication in works such as *Ethics in Rhetoric* and *Language is Sermonic*. In *Language is Sermonic* he asserts that all language use inescapably involves expressing some values, whether implicitly or explicitly. Even the supposed neutrality and value-indifference that scientists typically claim for themselves and their science is itself actually a value system, he points out, though a disguised one that denies that it is a value system. It is understandable, then, that the scientific mindset—and Weaver would have included the technological mindset as well—would try to deny that it is a value system. If it were to admit that it actually is a value system, Weaver explains, then it would lose its sense of absoluteness and primacy and erode its claim to privilege over mere opinion.

For Weaver, values serve as the foundation for rhetoric, giving it the substance toward which the audience is persuaded. Values come before our discourse rather than stemming from it, as the sophists held. Weaver took values very seriously and was a Platonist to the core. His essay on Plato's *Phaedrus* is still widely studied today as an explication of this dialogue from a perspective true to Plato's spirit.

Weaver was highly critical of the ascendancy of science and technology in our culture. He was concerned about the scientific and technological mindset for two reasons. First, it specifically distances itself from values, which Weaver thought must always serve as the motivating force behind our communications and actions. What we are left with, after removing values and the goals and emotional drives that go with them, is a strangely sterile discourse that encourages passivity and obscures the necessity for social discourse about topics such as opinions and goals. In this concern about the elimination by science of values as a motivating force for concrete action, Weaver is like Burke, whom he quotes in his famous essay on *Phaedrus*:

> Kenneth Burke . . . has pointed to "the pattern of embarrassment behind the contemporary ideal of a language that will best promote good action by entirely eliminating the element of exhortation or command. Insofar as such a project succeeded, its terms would involve a narrowing of circumference to a point where the principle of personal action is eliminated from language, so that an act would follow from it only as a non sequitur, a kind of humanitarian after-thought" (from Burke's *A Grammar of Motives*, 90, in Weaver, 22).

Rhetoric, Knowledge, and Values.   These notions have been elaborated recently in many different directions. In this book we can only glimpse these developments. The notion that science and technology represent value systems has flowered into an entire subfield within rhetoric studies, known as the rhetoric of science, with a parallel rhetoric of technology just now emerging. S. Michael Halloran, for instance, has shown the roles of rhetoric and implicit values in the process of determining who received credit for the scientific discovery of the double-helix structure of the DNA molecule. Dale Sullivan has shown how scientific discourse is saturated with a particular value system, making it very much like the ceremonial discourse that celebrates noble accomplishments. Scientific reports, for example, are credible and acceptable precisely to the extent that their research methodologies show scrupulous adherence to the prevailing standards—values—of investigation in that scientific community. Alan Gross and R. Allen Harris have shown how rhetoric works to negotiate dissent and build assent within scientific and technical groups. They have also shown how rhetorical argument is utilized in advancing entire new paradigms (or supertheories) within various fields of science. Craig Waddell has explored the role of values and power positions in the rhetorical negotiation of public policy decisions concerning science.

S. Michael Halloran and Merrill Whitburn together have shown the close connection between the rise of the plain style and the rise of modern science. The earliest modern scientists insisted on an austerely plain style that allowed the content, thinking, evidence, and argumentation of their discourse to come to the forefront. This has developed into the rhetoric of scientific discourse that we know today, characterized by distinct ideas developed by cogent arguments and supported by experimental evidence presented impersonally.

The notion that rhetoric inherently deals with values has developed in several directions, too. Stephen Toulmin has explained the inner workings and assumptions of real arguments (in contrast to abstract formal logic). He has found that objective, factual information is only one part of scientific and technical argumentation. Equally important are the "warrants" that justify drawing interpretations and conclusions from factual information. These warrants are, in effect, social conventions that have been repeatedly assented to and so are assumed not to need justification, Toulmin found. But these warrants vary from time to time, from culture to culture, and from community to community. Therefore one of the crucial elements of reasoned arguments is not absolute but instead is socially relative.

The notion that values and rhetoric are related to power and social dominance within a culture has found expression in the works of Henry Louis Gates Jr. Gates, though neither a rhetorical theorist nor a technical communicator, has revealed insights that are applicable to the technical world. Gates has taken a rhetorical approach to understanding discourse among African–Americans and between African–Americans and others. He found that power relations greatly affect the form that ethics and rhetoric take in any discourse. African–Americans developed an elaborate, highly useful system of rhetoric that is closely attuned to social circumstances but that differs considerably from the classical rhetoric found among non-African–Americans. The values communicated in these rhetorics differ likewise. For our purposes, Gates found that what counts as knowledge and what is accepted as persuasive arguments in the advancement of knowledge depends very much on the cultural context and on the power relations within that context.

Still stronger cases have been made in recent decades for the close relationship between ethics and rhetoric and for the relation between rhetoric, technology, and science. Michel Foucault, a French philosopher, has focused on the centrality of discourse in society generally. Foucault sees discourse as the broad collection of language use within a society, especially language use by the dominant social strata in dealing with important issues. This is the sort of language use that gets things done and justifies actions—and results in differences of power, status, and privilege. The influence of this collective language use is so profound that, in effect, it exists almost by itself in creating its own offspring. This "reproduction" of itself is often more important than the individual authors and speakers through which it acts. Paradoxically, according to Foucault we do not speak a language; rather, language "speaks" through us.

This language use reflects itself, basically, and perpetuates itself. It specifically does *not* reflect any absolute, external entities, and it does *not* facilitate open discussion because it shuts out potentially challenging inputs. Regarding ethics, for example, any supposed arguments about values within the community would actually only maintain the prevailing social order. The people making these arguments might *appear* to be free to criticize as they choose, but the fuller reality is that the language itself undercuts genuinely innovative speaking or thinking.

A second, related notion of Foucault's is that power, language use, and knowledge are closely interconnected. Any society and its hierarchy of different social levels is, in an important way, formed through language and perpetuated through continuing discourse. Our legal code, for example, is not only expressed in carefully chosen and precisely used language, but language use actually constitutes the law. New events are interpreted through this language use, and social changes are put into effect through it. In this way new developments are accommodated within the system, and the system perpetuates itself. This powerful influence of language use in shaping our thoughts and actions applies not only to such obvious activities as legal transgressions but also to the prevailing power structure and to knowledge. Thus the language use of the privileged class perpetuates itself while it discounts anything different. It values what it chooses to recognize while ignoring anything different.

Some technical communicators might not agree with the scope or force of Foucault's theory of language. It is true, though, that this theory goes a long way toward explaining how very difficult it can be to engage in truly free discourse aimed toward new ways of thinking. Sometimes what we are up against in our arguments is not just a single person or a single event but the entire weight of the whole social structure and its long history, which lie behind the particular person or event. This certainly was true regarding the technical communications associated with the *Challenger* disaster (which we will explore in a later chapter), but it is also true for the technical and scientific communications of many other organizations.

Although it is easy to see how his views might apply to the imprecise world of opinions and beliefs, such as in politics or the law, Foucault contends that his theory applies to scientific and technical knowledge, too. From his perspective, our traditional image of the scientist doing science is mistaken; instead, science does the scientist. Beverly Sauer, for example, echoes Foucault in having shown how in the governmental investigation of a mining disaster, the voices of technical and scientific experts were sought and highly valued. The voices of the wives of the male victims, however, were virtually ignored because they were not sufficiently technical and lacked the credentials of technical experts. The wives' awareness, derived from day-to-day familiarity with the miners, did not count as valid knowledge to the investigators, who were themselves technical experts. Sauer explains, however, that the nonexpert wives were aware of imminent danger in the mines well before any objective, technical indications became apparent.

Evelyn Fox Keller, the philosopher of science, has pointed out how thoroughly the scientific frame of mind has pervaded our thoughts and values. Keller has explained that though sex can be a useful notion in the scientific investigation of nonhuman organisms, it cannot be applied as usefully to humans. Indeed, the supposedly scientific perspective when applied to humans can become confusing and damaging in many ways. For that reason, Keller and many others prefer to speak of "gender" rather than "sex" in humans because "gender" carries with it the notions of flexibility, cultural conditions, and conscious deliberation. Keller also explains, however, that this distinction can cause its own problems because it deliberately distances itself from the scientific frame of mind, which our culture inappropriately thinks is the only basis for true knowledge.

The question of what should count as knowledge figures prominently in the thinking of many critical theorists in both the United States and Europe, including the German philosopher Jürgen Habermas. Habermas is concerned that the rise of science and technology so dominates modern culture that when we think of knowledge, we assume that it must be technical or scientific. Anything else seems not to count as important, including values, beliefs, and purposeful argument. About physical scientific knowledge, Habermas like Aristotle thinks that there is really nothing to argue about, and so it discourages public discourse. More important, this factual knowledge cannot in itself guide how we conduct our lives, whether individually or socially. Habermas is also concerned with the decline of public discourse that seems to have gone hand-in-hand with the rise of science and technology. If the only thing worth talking about is factual knowledge, which has only one side, then it is no wonder that public discourse is fading.

In response to these concerns, Habermas advocates recognizing that public discussion about values, goals, and policy amounts to another valid sort of knowledge. This sort of knowledge, however, develops through free, earnest discussions aimed at developing a consensus within a community, whether that community is a neighborhood, a nation, or a technical discipline. Only such discussions, not scientific or technical facts, can yield meaningful guidelines for deciding civic activities. These discussions are genuine arguments because no single, preexisting truth about such knowledge is to be found. Instead, a provisional truth must be determined within the particular social context. Not only is this consensual knowledge socially arrived at, but it also is not "known" once and for all because it springs from social interchange. This discourse never ends and is never finally closed. There simply are no absolute, independent criteria for deciding such knowledge other than rational argumentation among earnest, well-intentioned participants. Consensus is our only standard on such matters, and language is the medium that makes consensus possible. Thus Habermas calls for a renewal of public discourse focusing on values that will mitigate the dominance of factual knowledge in our culture.

The notion of science and technology as being embodied in discourse and perpetuated by it has been further developed by many others, such as Jean–Francois Lyotard. Lyotard considers science and technology as parallel pieces in a single grand story about the way the world is and the way we should live in it. This story or "master narrative" tells us, implicitly or explicitly, what counts as knowledge; where we should go to find answers to questions; which questions are important; what activities are important; and which people are important. It also tells us what and who are not important. Phrases such as "scientific progress" and "the advancement of science" illustrate this sense of participating in an ongoing grand story that accumulates as it moves toward an end. This end is obviously valued and can serve as a value itself when it justifies the means to achieve it. The work of scholars such as Lyotard and Foucault reveals to us this dynamic relationship among language, knowledge, and power, which otherwise would be inapparent because the narrative is taken for granted. This revelation in turn allows us to open up the narrative to reflective appraisal.

In conclusion, we see that historically, for the most part, ethics has been understood as inseparably connected with rhetoric and communication: We cannot have one without the other. Plato held that ethics comes before everything and that rhetoric is only a vehicle for the implementation of ethics in our lives. Aristotle held that ethical judgments have to be rhetorically argued in order to reveal the intrinsic goodness of the differing sides and to allow one to prevail. The sophists, on the other hand, contended that rhetoric and its techniques are quite separate from, even indifferent to, ethics. At best, ethics results from rhetoric. In more recent times, rationalists, social constructionists, and other theorists have shown that all language use entails ethical values and cannot do otherwise. We should, therefore, always be conscious of the ethical values communicated through the rhetoric of our discourse.

Topics for Papers and Discussion

1. Kenneth Burke links ethics to rhetoric by explaining how we cannot have one without the other. Though we often think of medical discourse as purely factual and objective, rhetoric and ethics play important roles even in this sort of technical communication. Read Mary Specker Stone's "In Search of Patient Agency in the Rhetoric of Diabetes Care." Stone uses Burke's notion of terministic screens and the pentad of dramatism to reveal how contemporary medical communication can become confused about who is doing what in what situation. As a result, a vitally important opportunity to persuade to effective action is missed, and the ethical responsibilities become confused.

First, write a brief report summarizing Stone's article. Second, suggest other ways in which values can become ambiguous through language use—or misuse—in other technical communications situations. In particular, consider the ways Burke's pentad can reveal aspects of a technical communication situation that otherwise would be hidden. Often in technical communication we focus only on what is being done and by what technical device but do not fully consider who is doing it and for what purposes. This second report can be done individually or in small groups.

2. It often appears to scholars in programs of Science, Technology, and Society and to others that technology seems to almost be a living organism with a mind of its own and a desperate will to preserve its own life and to grow. It seems that technology is pursued just for the sake of pursuing technology itself. Langon Winner, for example, has written about "autonomous technology" that seems to pursue a life of its own for no purpose other than its own continued existence. Jacques Ellul, Stephen Monsma, and many others have written similarly on ethical concerns about technology regardless of the particular form it might take.

The automobile is perhaps the best—or worst—example of this phenomenon. What began as basic transportation developed into farm machinery enhancing productivity and in turn enabling the migration of farmers to urban industrial centers. It also developed into a status symbol, a sport with many different forms, a vital mainstay of our economy, a liberating social device, a second home, an expression of personality, and many other "things" beyond just a device for basic transportation. In addition, it fostered an enormous petroleum industry and made us more dependent on the trade policies of other countries, just as it fostered an enormous civil engineering industry to plan, finance, build, maintain, and improve highways. The list can go on almost indefinitely. The point is that the "thing" as a technological artifact has come to mean much more than it did originally and to be valued much more and in many new ways.

Think of a similar example from history, the news, or your own knowledge. Research library sources to develop your example in depth. A number of scholarly, professional, and educational journals deal with topics of this kind at various levels; your library can direct you to more, some of which might even be available on-line in a full-text version. Among these are *Technology and Society; Science,*

Technology, and Human Values; IEEE Society and Technology; The Sciences (from the New York Academy of Sciences); and *Smithsonian* magazine. A particularly good source is Herman Tavani's bibliography, "Information Technology, Social Values, and Ethical Responsibility." Write a report and give a short oral presentation to the class or to your group on what you learned from this example.

3. One important new medium of technological communication is the World Wide Web. Well-crafted Web sites seem to have a rhetoric that persuades or leaves an intended impression on the viewer. Aristotle's rhetoric states that the three fundamental categories of rhetorical appeals are those based on logos or reasoned argument, on *ethos* or credibility based on character and involving ethical values, and on *pathos* or emotional appeals. Besides the textual content of a Web site, which would be an appeal based on logos, most sophisticated sites also involve features based on ethos and pathos. An appeal based on pathos would aim at the emotional reaction of the viewer, exciting, calming, or diverting the viewer in various ways. Obvious ways to do this involve colors, flashing images, and strong or evocative images.

Several Web sites I visited recently on the conflict in the former-Yugoslavia area, for example, portrayed many gruesome photographs of atrocities committed on each side. Evoking the strong emotional reaction of the audience was clearly a major purpose of this site. As a calmer example, the Web sites of a scholarly or scientific journal such as *The New England Journal of Medicine* also use color (or the lack of it) for both ethos and pathos. The background is simple white, like a page from a paper journal, with most of the display consisting of simple text. This supports an ethos of scholarly and scientific learning. The factual knowledge is the point, not the emotional arousal or the personal (rather than professional) opinions of the audience. The white color of the background and the near-total absence of other colored spaces or colorful images support the ethos of emotional detachment and rigorous learning. It also supports the disinterestedness of the physician in caring for any and all patients the same. Again, the information is the point rather than anything else. Ethos, remember, concerns credibility. The high reputation of this journal stems from credible, respected articles published in the past, and at the same time it projects credibility into the present and future. We should feel calm and secure under the care of any well-informed physician who would read this journal.

Search for one or two Web sites that seem to reflect clearly a concern for rhetorical impact, especially those concerning values, ethos, and pathos. Perform a rhetorical analysis on them explaining how these sites operate rhetorically. A fundamental assumption is that whatever features appear on the site are there deliberately and for a purpose. You might prefer to compare and contrast two sites rather than treating them individually. You might want to consult a basic work on rhetorical analysis and criticism such as *Rhetorical Criticism* by Sonja K. Foss. Also consult James H. Moor's "If Aristotle Were a Computing Professional" for insights on Aristotle's ethics in the computer age. This assignment could be done individually but would be even more productive and informative done in small groups.

4. Kenneth Burke's notion that we perceive and understand the world through terministic screens is one of his most intriguing and useful rhetorical contributions. Most of us are not used to thinking that things are defined by our language use. Instead we are used to thinking that things already exist "out there," separate from and before our thoughts or our language use.

Read Burke's essay *Terministic Screens* and discuss it in small groups to make sure everyone in the group has a basic grasp of it. To see how it applies to science, read also Mary B. Coney's *"Terministic Screens: A Burkean Reading of the Experimental Article."* Language is portrayed by Burke as a mixed blessing, both helping us to see and understand what otherwise would be a meaningless chaos of stimuli but also limiting our understanding by restricting us to a particular framework. This is true not only for literature, philosophy, and politics but also, according to Burke, for technology and science. Remember that scientific knowledge is grounded in specific theories that give meaning to observations, and these theories are articulated in language. Many contemporary thinkers believe that language, through theories expressed in language, even makes the observations themselves possible. These screens, furthermore, reflect values by selecting what we perceive.

Think of an instance from your own experience or the news that seems to show a terministic screen at work. You might, for example, have felt treated in a doctor's office as a patient or a clinical specimen rather than as a full human being. Or you might have felt treated as a case rather than as an individual human being by a government bureaucracy. Remember that Burke points out that terministic screens are like two-edged swords. They cut both ways; they can be used to protect or to attack. Think of both the advantages and the disadvantages of using the terministic screen in your example. Of course, it might make a difference whose screen it is and which side of the screen you happen to be on. The disadvantages might be easy to list but the advantages a bit harder though more enlightening. Usually this would involve delving into historical circumstances and complex economic and political issues. Remember too that terministic screens are usually unconscious devices—we are not continually aware of using them, or having chosen to use them, or even that they exist.

5. One of the concerns of Jürgen Habermas mentioned in this chapter was the effect that science can have in shutting off public discussion and debate. If a matter has been subjected to scientific investigation and results have been reported, Habermas says, often there seems to be nothing more to be said on the matter. It is as if because science has spoken, the truth is known and anything else said would be irrelevant or untrue. The same can be said for technology, Habermas would likely agree. In many of these instances, he would say, perhaps science and technology have been carried too far, or too much has been expected of them, or they have been inappropriately applied.

Think of some instance from the news or from your own knowledge and experience where this cutting off of further discussion seems to have occurred.

This could have to do either with technology or science. Transportation studies are notorious for being controversial, such as where to locate a new highway and how sound levels will be controlled. To the participants and the stakeholders involved, was the matter really settled to their satisfaction? Does this shutting off seem valid to you? Depending on the answers you have come up with, try to think of other instances in which participants were left with the opposite feeling. Sometimes, for example, those who have lost loved ones in a transportation accident want to know technically or scientifically exactly what happened. Other survivors in other accidents want just the opposite and would be completely uninterested in technical details.

Another example might be an investigation of the technical feasibility of exploiting some mineral resource. Often the determination that something is feasible seems to be interpreted as a warrant to proceed to enact what is feasible. It is as though facts are mistaken as values when *can be done* is mistaken for *should be done*. This criticism has often been leveled against the nuclear power industry and its proliferation of nuclear power plants in the 1960s and 1970s.

You might want to read further in the philosophy of Habermas to get a better sense of the depth and complexity of his concerns. A good place to start to familiarize yourself with Habermas in relation to technical communication would be in scholarly and professional journals. The references section of this book can direct you to basic writings by Habermas and articles about Habermas by Mathias Kettner, Nancy Roundy Blyler, and Helen Constaninides, which contain further references.

6. Advanced students might wish to delve further into the rhetoric of science.

On the other hand, the area of the rhetoric of technology is still in its formative stages, though investigation even in this limited area might be highly interesting. It could also provide students with the opportunity to conduct original investigation of publishable significance.

REFERENCES

Aristotle. *Rhetoric and Poetics.* Trans. W. Rhys Roberts (*Rhetoric*) and Ingram Bywater (*Poetics*). Ed. Friedrich Solmsen. Intro. E. P. J. Corbett, New York, New York: Modern Library, 1954.

Berger, Peter L. and Thomas Luckmann. *The Social Construction of Reality: A Treatise in the Sociology of Knowledge.* Garden City, New Jersey: Doubleday, 1966.

Blyler, Nancy Roundy. "Habermas, Empowerment, and Professional Discourse." *Technical Communication Quarterly* 3/2 (Spring 1994): 147–64.

Burke, Kenneth. "Terministic Screens." *Language as Symbolic Action.* Berkeley, California: University of California Press, 1966.

Coney, Mary B. *"Terministic Screens:* A Burkean Reading of the Experimental Article." *Journal of Technical Writing and Communication* 22/2 (1992): 149–58.

Constantinides, Helen. "Jürgen Habermas: 'What is Universal Pragmatics?' " *IEEE Transactions on Professional Communication* 41/2 (June 1998): 143–45.

Dombrowski, Paul M. "Can Ethics Be Technologized? Lessons from *Challenger*, Philosophy, and Rhetoric." *IEEE Transactions on Professional Communication* 38/3 (September 1995): 146–50.

Ellul, Jacques. *The Technological System*. Trans. Joachim Neugroschel. New York: Continuum, 1980.

Evernden, Neil. *The Social Creation of Nature*. Baltimore, Maryland: Johns Hopkins University Press, 1992.

Foss, Sonja K. *Rhetorical Criticism: Exploration and Practice*. Prospect Heights, Illinois: Waveland Press, 1989.

Foucault, Michel. "The Archeology of Knowledge." *The Archeology of Knowledge*. Trans. A. M. Sheridan. New York: Pantheon, 1972.

——— . *The Order of Things: An Archeology of the Human Sciences*. New York: Vintage Press, 1973.

Gates, Henry Louis Jr. "The Signifying Monkey and the Language of Signifyin(g): Rhetorical Difference and the Orders of Meaning." *The Signifying Monkey: A Theory of Afro–American Literary Criticism*. Oxford, UK, and New York: Oxford University Press, 1988.

Gorgias. "Encomium of Helen." Trans. Rosamund Kent Sprague. *The Older Sophists*. Ed. Rosamund Kent Sprague. Columbia, South Carolina: University of South Carolina Press, 1962.

Gross, Alan G. *The Rhetoric of Science*. Cambridge, Massachusetts: Harvard University Press, 1990.

Guthrie, W. K. C. *The Sophists*. Cambridge, England: Cambridge University Press, 1971.

Habermas, Jürgen. *Moral Consciousness and Communicative Action*. Cambridge, Massachusetts: MIT Press, 1989.

——— . "Discourse Ethics: Notes on a Program of Justification." *The Communicative Ethics Controversy*. Eds. S. Benhabib and F. Dallmayr. Cambridge, Massachusetts: MIT Press, 1990.

Halloran, S. Michael. "The Birth of Molecular Biology: An Essay in the Rhetorical Criticism of Scientific Discourse." *Rhetoric Review* 3 (1984): 70–83.

Halloran, S. Michael, and Merrill D. Whitburn. "Ciceronian Rhetoric and the Rise of Science: The Plain Style Reconsidered." *The Rhetorical Tradition and Modern Writing*. Ed. James J. Murphy. New York, New York: The Modern Language Association, 1982.

Harris, R. Allen. "Rhetoric of Science." *College English* 53/3 (1991): 282–307.

Jarratt, C. Susan. *Rereading the Sophists: Classical Rhetoric Refigured*. Carbondale, Illinois: Southern Illinois University Press, 1991.

Katz, Steven B. "The Ethic of Expediency: Classical Rhetoric, Technology, and the Holocaust." *College English* 54/3 (March 1992): 255–75.

Keller, Evelyn Fox. *Reflections on Gender and Science*. New Haven, Connecticut: Yale University Press, 1985.

Kennedy, George A. *Classical Rhetoric and Its Christian and Secular Tradition from Ancient to Modern Times*, 2nd, Chapel Hill, North Carolina: University of North Carolina Press, 1999.

Kerferd, G. B. *The Sophistic Movement*. Cambridge, England: Cambridge University Press, 1981.

Kettner, Matthias. "Scientific Knowledge, Discourse Ethics, and Consensus Formation in the Public Domain." *Applied Ethics: A Reader*. Eds. Earl R. Winkler and Jerrold R. Coombs. Oxford: Blackwell, 1993.

Lyotard, Jean–Francois. *The Postmodern Condition: A Report on Knowledge*. Trans. Geoff Bennington and Brian Massumi. Minneapolis: University of Minnesota Press, 1984.

Monsma, Stephen V. *Responsible Technology*. Grand Rapids, Michigan: W. B. Eerdmans, 1986.

Moor, James H. "If Aristotle Were a Computing Professional." *Computers and Society* 28 (September 1998): pp. 13–16.

Perelman, Chaim. "The New Rhetoric: A Theory of Practical Reasoning." *The New Rhetoric and the Humanities*. Dordrecht, Holland: D. Reidel Publishing, 1979.

Perelman, Chaim, and Lucie Olbrechts–Tyteca. *The New Rhetoric: A Treatise on Argumentation*. Trans. John Wilkinson and Purcell Weaver. Notre Dame, Indiana: Notre Dame University Press, 1969.

Rorty, Richard. *Objectivity, Relativism, and Truth*. Cambridge, England: Cambridge University Press, 1991.

Sauer, Beverly. "Sense and Sensibility in Technical Documentation: How Feminist Interpretation Strategies Can Save Lives in the Nation's Mines." *Journal of Business and Technical Communication* 7 (January 1993): 62–78.

Schiappa, Edward. *Protagoras and Logos*. Columbia, South Carolina: South Carolina University Press, 1991.

——— . "Sophistic Rhetoric: Oasis or Mirage?" *Rhetoric Review* 10/1 (Fall 1991): 5–18.

Stone, Mary Specker. "In Search of Patient Agency in the Rhetoric of Diabetes Care." *Technical Communication Quarterly* 6/2 (Spring 1997): 201–17.

Sullivan, Dale L. "The Epideictic Rhetoric of Science." *Journal of Business and Technical Communication* 5/3 (July 1991): 229–45.

Tavani, Herman T. "Information Technology, Social Values, and Ethical Responsibility." *IEEE Technology and Society Magazine* 16 (Summer 1998: 26–39.

Toulmin, Stephen. *The Uses of Argument*. Cambridge, UK: Cambridge University Press, 1964.

Waddell, Craig. "The Role of Pathos in the Decision-Making Process: A Study in the Rhetoric of Science Policy." *Quarterly Journal of Speech* 76/4 (1990): 381–400.

Weaver, Richard. "The *Phaedrus* and the Nature of Rhetoric." *The Ethics of Rhetoric*. Chicago: Regnery Gateway, 1953.

Weaver, Richard M. *Ethics of Rhetoric*. Chicago, Illinois: H. Regnery Co., 1953.

——— . *Language is Sermonic*. Baton Rouge, Louisiana: Louisiana State University Press, 1970.

Winner, Langdon. *Autonomous Technology*. Cambridge, Massachusetts: Massachusetts Institute of Technology Press, 1997.

CHAPTER 3

THE ETHICS TRADITION

In the previous chapter we reviewed the history of how ethics is linked inseparably to rhetoric as communication. In this chapter we review the history of ethics itself, focusing on the ethical theories of Aristotle, Kant, utilitarianism, and an ethic of care. These principal theories represent the broad span of the European–American tradition in ethics from classical times to the present. We briefly examine three other theories somewhat outside the mainstream of this tradition: those of Confucianism, Emmanuel Levinas, and Bernard Gert. Together they offer a range of approaches from philosophical to pragmatic and intuitive, with focuses on personal character or virtue, duty or obligation, utilitarian consequences, and caring relationships. The four principal perspectives represent the most highly regarded ethical approaches throughout European–American history, giving us the concepts and vocabulary we commonly use today. They were chosen not only because of this representativeness of common notions but also because they provide their own unique contributions to an overall, complex picture of ethics. Though other perspectives have their unique strengths, too, a survey of them all is beyond on the scope of this short book and might result in more confusion than clarity.

The perspective of Aristotle deals with virtue and personal character. It defines and explains basic notions such as goodness, truth, justice, and rightness as principles for guiding our conduct. It is a fairly pragmatic approach but with a philosophical cast. This cast imparts a sense that a given action should be performed principally because of its inherent goodness. It also lends a sense of having metaphysical principles and standards for guiding ethical determinations.

The perspective of Kant deals with duty or obligation based on a fundamental universal principle. This principle can be figured out rationally and so does not rely on metaphysical theories to support it. It explains that an action should be performed just because it is the right thing to do, regardless of its costs or benefits to us individually. It also strives for fairness and equality by showing that ethics can be understood by all people to apply equally to everybody.

The perspective of utilitarianism weighs the consequences of costs of an action against benefits in order to calculate the most socially desirable course of action. It is a calculus of sorts that strives to be fair (in the sense of not distinguishing between people) by being impersonal. It treats people somewhat like interchangeable parts of the social machinery and insists on being unresponsive to the interests or feelings of individuals. For these reasons we often find utilitarian calculations being performed by the government, trying constantly to be impartial and objective. The Federal Aviation Agency, for instance, weighs costs against benefits in deciding what policies to mandate, as does the Environmental Protection Agency.

The perspective of an ethic of care presents a new, nontraditional way of understanding ethics. This perspective, which was developed largely for the sake of being gender-sensitive, has the support of some feminists but not others, as well as the support of other ethical critics. Rather than insisting on impartial justice, for example, supporters of an ethic of care urge other standards for making ethical decisions, such as caring concern and the quality of relationships. They also urge flexibility and sensitivity to the particulars of a given situation rather than insisting on inflexible, universal rules.

The remaining three perspectives are not as fully discussed here as the others. They are presented to at least suggest the range of possibilities of important ethical perspectives that happen not to lie in the mainstream. Confucian ethics is Chinese and represents a culture other than European–American. Levinas espoused a very learned phenomenological and existential approach with strong theological undertones that is difficult to articulate because it is entirely opposed to the idea of systematizing. He also argued for a complete change of focus from the traditional "I" to "the other." Gert is an ethical theorist who deliberately does not diverge from the mainstream, instead reinterpreting traditional ethical concepts in contemporary terms for contemporary contexts and integrating them into a unitary system.

Ethics textbooks typically begin with a review of the history of ethical theories before moving to specific applications and cases. For many people nowadays, however, the traditional major figures in ethical theory no longer carry the authority they once did and so are often dismissed as irrelevant for today's world or at least impractical in their philosophizing. Though such judgments might be hasty, they reflect a genuine disinclination to philosophizing and theorizing among many people including technical communicators. For this reason, in this book we will review only a few of the principal ethical theories that traditionally have garnered enduring respect. No one theory will be advocated over any others. Instead the pluralistic and eclectic use of all theories, or even of new insights different from them, is recommended. Later, in the case studies, we will address ethical issues in technical communication in a flexible, eclectic way, drawing freely from these theories as they offer us useful insights. This will allow us to avoid prejudging these cases while affirming the responsibility and the ability of each of us to determine ethical issues for ourselves.

Let us first look at the major historical approaches to see how they still remain relevant for concrete technical communication situations in our own times. This will help us avoid hastily dismissing what highly respected thinkers have said on ethics and allow us to benefit from whatever insights they have to offer. Indeed, because ethical responsibility involves open, critical deliberations on particular issues, it would seem to be irresponsible to exclude out of hand anything that might help us in these deliberations by providing important insights that we might not have arrived at on our own.

The approach used here is much like that of Mark Wicclair and David Farkas in their groundbreaking 1984 article, one of the first to treat ethics in technical communication systematically. Wicclair and Farkas distinguish ethics from relativist, self-interested approaches, the law, and religion. They instead take a humanistic approach to ethics, one that aims at learning how to conduct oneself simply "as a good human being" (16). They focus on three types of ethical principles: goal-based (e.g., utilitarianism), duty-based (e.g., Kantian ethics), and rights-based. These types of principles are useful and relevant for technical communication because, Wicclair and Farkas explain, they "are uncontroversial and generally accepted" (17). Another similarity between our approach and that of Wicclair and Farkas is the underlying recognition that ethics entails a burden that is intrinsic to our human nature. It is also ultimately a personal responsibility, "a matter of individual conscience and will" (19).

ARISTOTLE

Aristotle's systematic analysis of ethics is an ancient perspective. But though it is ancient, it is not irrelevant to us; like the familiar but ancient Golden Rule, this perspective is time honored. It outlines the nature of ethics as a subject area, its basic issues, and the concepts and language that have shaped discussions of ethics ever since. (Of Aristotle's three treatises on ethics, only his *Nicomachean Ethics* is discussed here.) Reflective reasoning, prudent judgment, deliberate choice, conscious application of the will, and practical action are all involved in the notion of ethical conduct for Aristotle. Many of the thoughts articulated by Aristotle also occur, as we will see, in Kant as well as in more recent discussions about ethics. Even though Aristotle's thinking is often valued for being grounded in practical affairs, most of his writing is only an exploration of the general nature of the subject. They are theoretical works, though real examples are often provided. For that reason they do not offer a hierarchy of specific values, say, or a table of concrete prescriptions, for that was not his aim.

Aristotle's theory of ethics is used in this book in its own right and also as a sort of paradigm for many of the philosophical theories stemming from the classical period. Many critics understand Aristotle as making practical and concrete the abstract, metaphysical thinking of his teacher, Plato, who in turn gave philosophical legitimacy to his intellectual predecessor, Socrates.

During the medieval period Christianity dominated all intellectual thinking, and its ethical theories were strongly influenced by the classical thinking of Plato and Aristotle. Saint Augustine's (354–430 CE) theory was shaped by Plato and so showed a distinctly other-worldly, spiritual cast. It aimed toward saving one's soul so as to secure eternal happiness for it. Later Christian ethical thinking, particularly that of Saint Thomas Aquinas (1225–1274 CE), was shaped more by Aristotle's practical ethical theory that aimed at achieving the best conduct in our temporal lives in the here and now. It assumes that guidance for practical conduct can be derived from first principles about human nature and from a reasoned understanding of the interrelation of the physical and the metaphysical. It is fair to say that for Aristotle ethical principles could be derived from analytical observations about human affairs, which is the source of theories. Despite its practical bent, metaphysics still plays a role in Aristotle's ethics, though rather indirectly. Transcendent, metaphysical truths lie behind our observations and so still serve, though much less directly than in Plato, as the ultimate foundation of ethics.

Ethics, Aristotle says, is the study of what is involved in good actions. He begins by explaining the nature of this entire subject. This methodology, which first establishes the nature of the subject, gives Aristotle's theorizing a feeling of sensible naturalness. Ethics is a subject that simply does not allow hard and fast answers, as mathematics does, Aristotle says. Every class of things has an inherent degree of precision, and, he says, "it is the mark of an educated man [person] to look for precision in each class of things just so far as the nature of the subject admits" (329). This simple statement applies as well to our modern concerns, indicating that we should not try to make ethics more definite than it can be. For example, a code of ethics cannot but be general, without concrete directives that can be applied without lengthy interpretation to particular cases. We cannot therefore expect to create a technology of ethics, a point that is developed more fully elsewhere (Dombrowski, "Can Ethics Be Technologized?"). Though some scholars have attempted to create a mechanical formula for making decisions, for the most part these approaches have not found wide acceptance (e.g., Sturges).

Furthermore, when the definiteness of technological matters is coupled with the indefiniteness of ethical matters—as in the ethical study of technical communication—any inclination we might feel to technologize is intensified by the very positioning together of these two quite different matters. We might naturally feel a sort of frustration that the clear definiteness in one area cannot also be found in the other area. And, people being what they are, they try to avoid frustration. Thus we might feel an urge to avoid considering the ethical dimensions of technical communication. We should, however, resist the inclination.

Ethics, Aristotle reasons, is about what is sought for its own sake—goodness itself—and not for the sake of something else such as money or success. Though this statement might at first strike us as idealistic and impossibly abstract, it does in fact correspond to our everyday understanding of the sort of behavior we admire as ethical. When we say someone behaved responsibly or ethically, we usually mean that that person did the right thing. "The right thing" here means not what was easy or typical but what somehow went beyond that in doing more than what was expected, easy or understandable. And doing it just *because* it was the right thing to do—that is, for its own sake.

Aristotle, though deliberately abstract (for that is his methodology, to abstract theory from practical experience by applying reasoning), recognizes the complexity and contingency of our human condition. Humans are uniquely compound creatures, having aspects of both a lower, animal nature and a higher, divine nature, Aristotle says. We have instinctual drives and appetites such as found in animals, which must be satisfied yet which have nothing to do with goodness. We should, however, seek and act out of what we share with the divine even though we are fated to live out our lives in the sphere of the mundane and imperfect. What we share with the divine is our rational powers, which we must cultivate carefully.

Interestingly, ethics for Aristotle involves the disposition to seek after the good, and only to that extent can we really be virtuous. It is the person and not the action that is virtuous. Virtue has nothing to do with single, specific actions because an action might be performed for all sorts of reasons, few of which might be innately virtuous. Instead, virtue as ethical behavior is a habit that must be cultivated so as to yield the disposition to behave in virtuous ways, to choose and to will ethically. (A modern social psychologist would call this habitual disposition an attitude.) Though we are creatures of habit, from Aristotle's perspective we are ethically responsible for our habits. We become habitually virtuous by deliberating about the ethical and then behaving accordingly, the repeated actions shaping our habits (and not the other way around, as we usually suppose today).

G. E. R. Lloyd, a famous Aristotelian scholar, says that one of Aristotle's key contributions to the theory of ethics is his assertion that each person is responsible for his or her character, which determines the goals the individual desires and acts toward. "The self-indulgent man may be held responsible for becoming self-indulgent, for his character is the result of repeated voluntary acts of self-indulgence, and so too the man who is unjust is responsible for having become so," Lloyd comments (231).

Such thinking might at first strike us as needlessly indefinite and abstract, but it does in fact correspond to our everyday understanding of the ethical, even as it relates to technical communication. If you, for example, were to behave solely in the ways dictated by your circumstances, say according to the rules and regulations of your corporation, this would not, from Aristotle's perspective, necessarily constitute ethical behavior. You are doing only what you are told to do because you are required to do it and not for the intrinsic rightness of the behavior. Or, more pragmatically, you are doing it for the sake of retaining your job and securing your financial condition, which again would not be for the intrinsic rightness of the behavior. Some modern thinkers hold a somewhat more complex view, though, namely that one has a major ethical responsibility to make a decent living for oneself and one's family, in addition to other responsibilities (e.g., Markel).

Ethics for Aristotle derives from reason, and so ethical behavior must be reasoned behavior. What this means practically is that ethical conduct is not automatic or unreflective. Aristotle explains that we usually do not especially admire someone for doing what comes so easily to that person that it is hard for them to do otherwise. Aristotle repeatedly refers to fortunate circumstances as making apparently virtuous behavior easy and almost unintended and so not really virtuous. This is one reason that he focuses on cultivating the disposition of the person rather than the behavior itself. We, for instance, do not admire as particularly virtuous a rich person at a checkout counter who returns the surplus if given too much money as change, although we would if the person were poor.

In other ways, though, Aristotle's ethics might differ from our own. Ethical behavior, he says, must also be deliberated, actively weighed, and specifically chosen for its own sake. It also must be done from a firm, unshakable character that was deliberately shaped over a period of time. This results in a virtuous character. He would not say one is ethical or virtuous when one performs just a single action but only as one's character is reflected in a continuous pattern of behavior over time. To behave ethically only occasionally, on the other hand, would, from Aristotle's perspective, only raise a question about the motives behind the actions.

One is not ethical or virtuous instinctively, Aristotle further explains, or genetically, as we might say today. To behave as our drives lead us is not to be led by reason itself. Furthermore, we would only be doing what is easy for us, not what is difficult, problematic, or troublesome. Most people nowadays think differently, of course, that one can indeed behave ethically occasionally, though most of the rest of the time one might just go with the flow, taking the easy path and following conventional norms because one is not confronted with compelling ethical dilemmas.

The chief strength of Aristotle's perspective for our purposes is that it is not idealistic. Though our power of reasoning constitutes our participation in the divine for Aristotle, we are nevertheless creatures burdened with deciding how to act in a pragmatic, imperfect world. Because of this fact of life, Aristotle avoids the extreme of declaring absolute rules, insisting instead on a mean between extremes as the guide for our actions. We should not be led only by feelings such as anger, or pity, or pain, or pleasure, but we must always struggle to judge and weigh issues through reason. We must deliberately decide our actions, he says, "For in everything it is no easy task to find the middle . . . but to do this to the right person, to the right extent, at the right time, with the right motive, and in the right way, *that* is not for every one, nor is it easy; wherefore goodness is both rare and laudable and noble" (368).

Aristotle's ethics is unusual in not being metaphysical or idealistic as, say, Plato's. It is firmly connected to the practical. Knowledge of morality and ethics is necessarily imprecise, for example. Guthrie concisely contrasts the practical Aristotle to the idealistic Plato. For Aristotle, "The object of our inquiry is not to know what virtue is, but to become good men," whereas for Plato, "to know what virtue is" is a prerequisite of becoming good (54). Scientific knowledge, by contrast, can be very precise and certain because it deals with what is entirely necessary and cannot be otherwise. We know that apples fall to the ground, and we cannot meaningfully conceive of the realistic possibility of their moving otherwise. Likewise for technical knowledge, for the same reasons. The modulus of elasticity of a bar of steel, for example, is definite and cannot be otherwise. For this reason we do not deliberate about matters of empirical necessity, that is, matters of science and technology, Aristotle says. Ethics, on the other hand, is a very different matter because it is concerned with the possibility of otherwise. A range of possible behaviors must always exist before ethics comes into play. That is, ethics is the problematic. So, from Aristotle's perspective, ethics in technical communication would not have to do with technology itself but rather with our decisions about how and when to use it.[1]

Although Aristotle examines ethics theoretically, his examination is not impractical. Instead, he insists, moral wisdom must be combined with practical wisdom in order to yield ethical action. Besides knowing abstractly some of the principles that should guide conduct, one also needs to understand fully the practical situation in order to deliberate how to act in a given situation.

> The man who is without qualification good at deliberating is the man who is capable of aiming in accordance with calculation at the best for man of things attainable by action. Nor is practical wisdom concerned with universals only—it must also recognize the particulars; for it is practice and practice is concerned with particulars (Aristotle, 460).[2]

Although Aristotle is often thought of as excessively cerebral and coolly rational, we should recognize that he also often expresses his thoughts in very homely illustrations that anybody can relate to. In explaining the nature of the various impulses and emotions that people feel and the relation of these to virtue, Aristotle explains that the highest principles find expression in the immediate, material realm. For that reason Aristotle explores in depth the nature of friendship and its relation to the principle of justice, for example. He does the same for love, which is its own motivation, just as with moral goodness. He explains that the best, highest form of love is specifically without concern for material gain or even reciprocation. Instead, it delights in loving for its own sake and for nothing else. Mothers, Aristotle explains, "love their children even if these owing to their ignorance give them nothing of a mother's due" in return (515).

This analysis also holds for friendship. A friend, he says, "wishes and does what is good, or seems so, for the sake of his friend" (535). In both these cases, love and friendship, the common denominator is the abandonment of self-interest or material gain, which is also the basis of morality and ethics. So, for Aristotle, despite his rigorous reasoning and formidable abstractive powers, ethics remains relational. It consists of how one relates to oneself, to the divine through the powers of reasoning, and to one's fellow human beings.

---

[1]This position differs considerably from that of some modern thinkers, such as Ellul, Monsma, and Winner, who contend that the pursuit of technology for its own sake, wherever it might lead, is in itself a choice of values and therefore an ethical matter.
[2]Both Dale Sullivan and Carolyn Miller make similar points about the complex uniqueness of real ethical dilemmas.

In our own times, as we will see later, renewed interest in ethics has been motivated in part by a desire for more caring and less self-interested relations between people. This is coupled with a recognition of the intrinsic validity and worth of relationships, simply and solely for the other person's sake. We will find this, for example, in Nel Noddings' ethics of caring and in Emmanuel Levinas' ethics of "the other," to be discussed later.

If ethics often requires facing difficulties, this is not to say that the only truly ethical behavior is that which brings hardship and suffering to us, of course. Aristotle goes to great lengths to relate ethics to all manner of human activity as it is informed by reasoning. Specifically, he draws strong connections to law and to politics, areas that in their best form must be ethical in that they concern what is best for human society. Good laws prohibit unjust behavior and so are linked to moral, ethical principles. Likewise, a good political society is structured and operates to ensure, to the extent practical, ethical social interactions. Nonetheless, Aristotle is quick to make clear that ethics cannot be reduced to politics or the law because it must guide us when the law or political rules are silent or in error. In two of the real cases we will examine—the *Challenger* disaster and Nazi technical information—ethical considerations specifically have priority over the prevailing legal and political circumstances.

Though one must always be careful to understand fully one's circumstances and the likely consequences of one's behavior, there are limits to these pragmatic concerns. Aristotle is clear that we should resist some deeds regardless of the personal consequences they entail. Applying this to our own times, we know that certain people have an ethical obligation to point out as strenuously as possible serious dangers stemming from technical matters. In fact, this obligation specifically in opposition to practical concerns (such as pleasing an employer, conforming to the accepted corporate expectations, or retaining one's employment) has been institutionalized in what we commonly call "whistle-blowing" laws.

As we will see, the *Challenger* disaster is a clear example both of the importance of whistle blowing and of the serious, practical consequences of doing so. Roger Boisjoly, author of the famous "smoking gun" memo cited in numerous technical writing textbooks, indeed lost his position because of his strenuous objections to certain organizational procedures and decisions. Threatening to file suit against his employer under the whistle-blowing laws, he was reinstated in his previous position after having been threatened with being marginalized and ostracized. It was a hollow victory, however, because his employer made his work environment so unpleasant that Boisjoly voluntarily resigned from his job. Thus the employer's original, punishing aim was accomplished regardless of the law. In the case of Star Wars, we will see that David Parnas insists that the American public has a right to expect truthfulness from those managing public funds and effectiveness and efficiency in the way their money is spent.

Aristotle's practical dimension is crucial to us. To engage in practical activity in the social sphere is typically human, and this sphere naturally includes the political, economic, industrial, and corporate arenas. In this way Aristotle pointedly distinguishes himself from the idealism of Plato and Socrates. Plato's pursuit of reasoning alone, in contemplation, was impractical. More importantly Plato's idealism abandoned a sense of realistic responsibilities in the messy give-and-take of the everyday world in which the vast majority of mankind lives.

But how can we relate Plato's and Aristotle's searching after the true, good, and right to our contemporary technical communication? Certainly their language seems out of place in the modern world of technology and science. Such language seems moralistic when applied to technology and science largely because we assume it to be inappropriate. Technology and science, we are used to thinking, deal with cold, hard facts that are indifferent to our feelings or ethics. That we would think this way automatically is, surprisingly, partly a reflection of the powerful influence of Aristotelian thinking on the Western intellectual heritage. Remember, Aristotle explicitly distinguished science—which deals with absolutely true, certain, and unchangeable knowledge—from rhetoric and ethics, which deal with matters of opinion and so are always uncertain and open to alteration.

Many modern thinkers differ from Aristotle's view that specialized knowledge such as in technology and science is separable from ethics. They feel that technology and science actually do embody systems of values and so represent a certain opinion or frame of mind about what is important, good, or right. The true, right, and good are highly operative in technology and science, they contend. Technology and science aim at the progressive refinement of knowledge through empirical observation and hypothesis testing. This is a rigorous, scrupulous process that aims toward the truth, even if it never fully attains it. That is why, for example, we turn to the scientific method in epidemiological studies in order to find out what really accounts for a disease rather than relying on unconfirmed personal opinion. That is how we know that it is safe to work and live with AIDS patients if certain behaviors are avoided, even though common opinion held that proximity itself was practically a guarantee of infection with HIV.

The scientific method, which applies to technological investigations as well, also reveals how science pursues the right. By strenuous negotiation among scientists and rigorous demonstration over the years, we have come to acknowledge that the method, if properly applied, must yield knowledge of a fairly certain sort that cannot be wished away. Therefore, we have come to prize the scrupulous application of this method as the *right* means to follow to reveal scientific and technical truths. The results are right to the extent that the method was followed conscientiously, and following it is the right thing to do because new reliable, practical knowledge will follow from it.

The pursuit of the good is also operative in science and technology. For example, clearer, faster, and more useful communication is commonly thought to be good for technical communicators. New digital technologies are assumed to be a good in themselves, for example.

Although it is interesting to see that the methods of science and technology amount to a pursuit of the true, good, and right, it should not surprise us. In recent years, more and more intellectual critics are coming to recognize that a paradoxical fuller truth lies behind misleading appearances. They recognize that technology and science are the genuine inheritors of the ancient philosophical tradition of creating new knowledge through rhetorical negotiation.

Richard Rorty, for instance, redefines the history of philosophy and science. He explains that while philosophy has traditionally been understood as seeking the absolute, real truth that presumably lies behind subjective appearances, contemporary science is also trying to do exactly the same thing (21–34). By carefully observing apples falling from trees, we learn the inapparent yet very real truth of the laws of universal gravitation. For this reason, Rorty feels that science actually is a form of idealism. Likewise Bruce Kimball explains that the tradition that sees philosophers as the creators of truths continues in our own times, but now in the form of scientists (175–221). Scientists, like ancient philosophers, distance themselves from everyday appearances and personal opinion in order to pursue fuller truths that are hidden to lay people. Thomas M. Lessl writes similarly but goes further in calling scientists the religious priests and philosophical oracles of our day. Their methods are hidden to us and distant from our everyday experience; we turn to them for truths that evade everyday perceptions; and we receive their revelations as the absolute truth, Lessl points out. He also says that their communications both to lay audiences and to scientists themselves have an exhortative, mystical component that clashes with the concrete realism of the knowledge they pass on. Carl Sagan and Jacques Cousteau, for example, saw themselves as on missions not only to communicate knowledge but also to propagate faith in the methods and aims of science. Their aims amount to a sort of salvation that seeks to fully understand our proper place in a sublimely wondrous creation. Furthermore, they speak not so much as unique individuals but to and for us almost as truer images of ourselves, as religious priests have done throughout the ages.

KANT

Immanuel Kant is perhaps the most important European philosopher of the period from classical Greece to the twentieth century. His ethical theory is based on a sense of duty, which is important both for what it is and for what it is not. As a philosophical position it avoids articulating a complex system of values and weighing particular contingencies. It also avoids understanding ethics as mystical dictates handed down to us from a higher authority, such as we often find in theological discussions of ethics or in spiritual philosophies such as Plato's. In addition, at a very practical level, Kant explicitly and specifically avoids both the circumstantial contingencies of relativism and the weighing of competing interests such as those found in utilitarianism and several recent ethical theories.

For Kant, self-interest, material gain, and the greatest good for the greatest number are all beside the point. His ethics is a deontology, an ethical system emphasizing obligation or duty. It is based entirely on binding, absolute duty and obligation as they guide the application of a free will in executing what Kant calls the universal "categorical imperative" rule of ethics. Though the particular course of action for any given real case is left undefined by his theory, the universal rule itself is hard and fast, neither suggestive nor conditional. It has nothing whatever to do with purposes or results and nothing whatever to do with feelings or emotions, including any sense of "moral feeling." In this way, we will see, it differs from both utilitarian ethics and an ethic of care.

Kant's language, admittedly, can impede an easy reading of his theory. The term "categorical imperative" itself has a tone that might make us shudder. In a way it should, but in another way it should not. Certainly in our own times, over two hundred years after Kant, we have become much less receptive to any claims to directive authority such as denoted in "imperative." This is due not only to our democratic, egalitarian social order but also to the broad intellectual movement known as postmodernism, which pointedly rejects authority and traditional foundations.

We are also put off by "categorical," which sounds so pointedly absolute, inflexible, and unconditional that it hardly seems possible for modern minds to take seriously. We have so cultivated our skills in critical reasoning and in searching for complexity and qualifications that the entire notion of the categorical can seem foreign: We know never to say never, for example. Any basic logic textbook, for instance, is likely to treat categorical propositions more as examples of fallacious overstatements than as exemplary thinking. Yet, from Kant's perspective, the fundamental principle of ethics cannot be other than categorical and imperative.

All humans are endowed with a sense of moral reason, he explains. Though we might at first glance take this to be an intuitive feeling for what is right and wrong, that would be incorrect. Kant is scrupulously careful not to sentimentalize ethics and our moral sense but rather to insist that ethics can be understood and derived only from abstract reasoning itself. Only reasoning, he says, can grasp the unseen but metaphysically real principles that drive our actions. Our senses, on the other hand, are too directly engaged with shifting appearances, confusing contingencies, and material satisfactions to allow them to derive any knowledge about our moral obligations, which are necessarily metaphysical.

In this way Kant differs considerably from, for example, the ethic of care advanced by some feminist theorists and others. An ethic of care commonly asserts the centrality of intuition and feelings, as we will see more fully later. For instance, Evelyn Fox Keller's celebration of the scientific contributions of a long-neglected biologist, Barbara McClintock, was titled *A Feeling for the Organism*, reflecting its concern with intuitive feelings about and appreciation for the whole organism. Such feelings have traditionally been strenuously avoided by scientists as a subjective contaminant of objectivity, feminist critics and others contend.

Kant's explanation in *The Fundamental Principles of the Metaphysics of Ethics* leading up to the principle itself is very involved and lies beyond our scope here. Suffice it to say that Kant's ethics is based on one's freely chosen decisions to act in good will out of a sense of duty. He derives this principle solely from close reasoning about the nature of humans and of reason itself. Our distinguishing feature as human beings is our reasoning abilities, according to Kant, which must therefore serve as a basis for judging ethics. Because of the presumed universal and absolute nature of reason, ethics, which derives from reason, must itself also be universal and absolute—that is, categorical. Entailed in the absoluteness of reason is the force of its determinations, which are not just suggestions or idealistic wishes but hard and fast directives commanding our compliance with its law. From this obligatory nature of ethics derives the fundamental importance of duty as the responsibility to carry out these directives regardless of anything else, solely because it is the right thing to do.

His fundamental ethical principle, the categorical imperative, is this: "Act as if the maxim of your action were to become by your will a universal Law of Nature." We can restate this as: Act in such a way that, if you had your way, the principle guiding your actions would become a universally binding law that everyone must act in accordance with (including in relation to you), applying to everyone, everywhere, and always, without exception.

The categorical imperative is a principle that, though it has the force of a law, in a paradoxical yet highly significant way it also does not. Usually we think of a law as something imposed on us by others, with which we are expected to comply and for which we will be punished if we do not. The categorical imperative, however, "must be looked upon also as giving itself the law" because it derives from the reasoning power that all of us hold in common (33). So the entire basis for the law is oneself in one's own nature. But, because this nature is common to all other humans, we should all come to the same rational recognition of the imperative equally and independently. In addition, the principle is universal in the breadth of its supposition, namely that the maxim guiding the action would, rationally, be willed by us as a *universal* law. Thus, the self, paradoxically, is also a universally legislating will. Furthermore, this act of universally legislating binds oneself in a way that is not a rankling imposition but a reasoned recognition of how things must be—they cannot rationally be otherwise. Therein lies the true meaning of the sense of duty that Kant emphasizes, not as a response to a command or to a system of rewards and punishments but as a conscious recognition of one's obligation.

Here, too, I might add, another apparent yet meaningful paradox occurs in Kant. The peculiar nature of this categorical imperative, which is not exactly a law in the usual sense, is that it requires recognition and acceptance by oneself as a binding law, consistent with its springing from reason. Every person, that is, would arrive at the same recognition independently without coercion or contingencies and must personally will it to be so in his or her life. Only by assuming the complete autonomy of each individual human will can one "act so that the will can consider itself at the same time as legislating universally by means of its maxims" (52). We usually think of a duty as an onerous burden and a limitation of our freedom. But, paradoxically, Kant's theory of duty is founded on the radically autonomous free will and its capacity to choose otherwise and on its reasoned self-persuasion not to choose otherwise. It is duty based in freedom.

This, Kant says, is one of the most important contributions of his theory to the history of ethics. He asserts a radically autonomous free will bound by duty, coupled by reason with a radical individuality that is nevertheless one with the universal. In this assumption of the centrality of reason to human beings, Kant deflates if not negates some of the difficult issues that commonly arise in ethics studies: the self versus society, free will versus lawfulness, natural impulse versus deliberated action.

Because of the metaphysical nature of ethics and of reasoning and because our understanding of ethics is derived purely from reasoning, ethics must be fundamentally disengaged from the physical world. One implication of this awareness is Kant's insistence that our decision to guide our behavior in accordance with the categorical imperative must be absolutely disinterested in any rewards or resultants or even any feeling of self-satisfaction we might have. Kant, surprisingly perhaps, insists that we not judge ethical matters on the basis of good feelings or of good outcomes, such as how many people will benefit from our decisions. "The moral worth of an action then does not lie in the effect which is expected of it" (16). In this way, Kant's theory is diametrically opposed to consequentialist ethical theories such as utilitarianism.

Kant also examines the nature of the individual in relation to society. Interestingly, because he derives the categorical imperative from the principles of reasoning that characterize us as human beings and that we all share in common, the same principle must apply to all, be valid for all, and be understood by all in the same way. Thus, though personal conduct in particular situations is decided and willed by the individual, it ought to be on the basis of the categorical imperative that, by definition, should guide everyone in the same situation similarly. In this way the personal is melded with the universal, and individual conduct is prevented from becoming arbitrary or relativistic. The individual acts responsibly as a rational being, as a single particular instance of a universal way of being.

Kant understands each person, as a rational being, to be a fundamentally autonomous self-legislator, making, willing, and enforcing laws for oneself about how one will behave. But this behavior, due to the nature of reasoning itself, must be in accordance with the categorical imperative, which must in turn necessarily be identically understood by all rational beings. Thus one's ethical decisions, though autonomous, are never egocentric or arbitrary or even self-serving but are always conforming freely and deliberately with a universal principle that cannot be evaded. "[M]an is subject only to his own and yet to universal legislation, and . . . he is obligated to act only in accordance with his own will which, however, in view of the end of nature is a universal legislating will" (51).

This is more than conformity, though, for it is done entirely for the sake of the principle itself. This is how the value of one's character shows itself, not from habitual inclination, or from intention, or from expectation of reward but solely from a sense of duty. In this way, we should note, the traditional dichotomy between the individual and society, so often found in other ethical viewpoints, is nullified in Kant as the individual is melded with all people acting rationally—as a single case of the universal.

The language, the deep metaphysical reasoning, and even some of the sentiments of Kant's ethical theory might seem strange to our modern tastes. It nonetheless has a good deal of relevance to understanding the ethics of technical communication. First, it emphasizes a sense of duty, doing what is right regardless of competing interests or eventual outcomes. Second, it conceptualizes ethics as both an individual and a social matter, or, more precisely, it defines one's personal ethical responsibilities in terms of a generic universal human being. Third, though this simplifies it, Kant's theory amounts for all practical purposes to the Golden Rule. In insisting that we should behave in ways such that all other people would want to treat others and also to be treated by others, Kant is entirely congruent with the proverbial Golden Rule: Do unto others as you would have them do unto you. He derives his principle, however, not from received, traditional authority (i.e., religious texts and authorized interpreters) but from rigorously reasoned argument from first principles. Kant assumes nothing except that we are all rational beings.

Of course, this assumption implies a great deal, namely that we are all capable of reflecting on our consciousness, of reasoning toward binding universal conclusions, of weighing judgments about practical action, and of willing our actions to conform to our judgments. Note that this assumption implies that we are roughly equivalent, at least insofar as reasoning is concerned. We all have the innate reasoning faculty that allows us to perform the ethical tasks we are burdened with, but, more important, we should all arrive at the same ethical appraisal of the same given circumstances. Kant's is decidedly not, then, a relativistic sort of ethics.

Still, we need to keep in mind that though the ethical force of the imperative is strong and clear in binding us to our duty, the particulars in any real ethical dilemma are rarely strong and clear. Despite the directive tone of Kant's statement, we still need to decide to undertake the carefully reasoned weighing of our ethical situation and then do the best that we can—"best" meaning reasonably such as we would expect others to do out of conscientious sense of duty in the same situation.

Perhaps this lends a sort of consolation for the awful burden of having to come to one's own judgments. We need not think in terms so much of, What do I really think or feel about this? or What do I really want to do? but in terms of what any other reasonable person would decide in such circumstances and how I would wish to be treated by others were I involved in this ethical dilemma. Thinking in terms of universalities or in terms of oneness with the rest of humankind can for some people be a relief. For these, thinking of the *categoricalness* of Kant's ethical imperative would be helpful. For others, though, it is not a relief. For these, thinking rather in terms of *imperative* duty regardless of what others might decide in a similar circumstance would be helpful. In either event, as generic human or as single individual, Kant's ethical theory can be similarly meaningful.

An interesting and directly relevant implication of the metaphysical nature of Kant's theory is its deliberate, radical distancing from empirical experience. The world and our actions in the world are observed through our senses. Our senses, however, can "never . . . fully discover the hidden springs of action, because examination of moral values does not depend upon the actions that one sees, but upon their inner principles which one does not see" (23). Kant goes so far as to insist, "One could not serve morality worse than to derive it from examples" (24). So when Kant offers concrete examples, they are not as foundational bases from which ethics should be derived or inferred but only as feasible illustrations of how ethical deliberations might be conducted and appraised. This, we should keep in mind, is the same reason for the cases presented here in this book.

Some of Kant's examples are simple and practical, surprisingly so for the abstruseness of his reasoning. They are quite relevant for technical communication. One example considers whether a merchant should take advantage of an inexperienced customer. Kant notes that where business is thriving, no merchant would take advantage of a customer because that would negatively impact the merchant's business in the long run. But the merchant's behavior, though it might commonly be called "honest," cannot be judged at face value, that is, on the basis of the action itself. Remember that for Kant, the principle guiding one's action—not the action itself or the results of the action—determines its goodness. Kant says that we would need more information before we could determine whether the merchant was behaving out of a sense of duty rather than out of material self-interest.

It is not a great leap from this example to technical communication consulting, for instance. Often consultants are in a position of greater knowledge and experience than their clients and easily able to take unfair advantage of them. Overcharging for time spent consulting on a project is one way to do this. Foisting plagiarized or pirated material as one's own is another way. Though we know that in the long run this deception would likely be detected and so we avoid doing it, from Kant's perspective this would not be done out a sense of ethical duty but only out of simple prudence and self-interest.

Regarding technical arts specifically, the subject matter of technical communication, Kant considers these to be ethically indifferent. "All sciences [and technologies] have a practical part consisting of problems stating the possibility of some purpose, and of imperatives that direct how to realize that purpose. The latter may therefore be called in general imperatives of skill. Whether the end is reasonable and good is of no concern here at all; but only what must be done to attain it" (35–36). Kant draws on a simple, practical example to illustrate his point. "The prescription given by a physician in order to effect the thorough cure of his patient, and that prepared by a poisoner to bring about certain death, are both of equal value in so far as each serves to realize its purpose perfectly" (36). The technical excellence of these actions is high in being both very effective and expedient, but certainly one is ethical and the other not.

Though the physician–poisoner example might strike us as too obvious, it is not far off the mark as to one of the real cases we will discuss more fully later. Steven Katz, in his article on technical communications about Nazi human extermination technology presented in one of our cases, points out that technical excellence or expediency cannot serve as a basis for rendering ethical determinations. The technology and the technical communications about the technology were effective and expedient but also appallingly unethical. From Kant's perspective, the principles or motives driving the actions would be crucial for judging these actions. Katz clearly would agree with Kant.

Kant's theory has even stronger relevance to technical communication. He explains that in order to put this metaphysical principle, the categorical imperative, into practical realization, it must be based on something "the existence of which of itself has an absolute value, which serves as an end in itself" (45). And this something is humankind because humans are rational, and our rational nature exists for itself. So for Kant every person must be taken as an end in itself and never as a means to an end. This leads Kant to another restatement of the categorical imperative: "Act so that in your own person as well as in the person of every other you are treating mankind also as an end, never merely as a means" (47).

To illustrate this point, Kant uses the example of a simple lie in making a promise to another person. In lying, one uses the other person as a means to one's own ends. This is fundamentally unethical, "For the person whom I am about to use for my purposes by such a promise cannot possibly agree with my conduct toward him" (47). If true for lying, it is infinitely more true for the several examples from Nazi technology that we will examine later. In all these cases, one's behavior to others was as a means to an end, whether toward the supposed purification of a master race or toward the claimed advancement of scientific knowledge.

In contrast to Kant's theory derived from abstract reasoning and insisting on duty, a quite different approach to ethics is the ethic of care proposed by some feminist thinkers and others recently. It will be discussed more fully later, but suffice it to say here that some feminist critics object to traditional thinking about ethics. They consider both an exclusive reliance on abstract reasoning and an insistence on rigidly following impersonal directives to be a distinctly masculinist approach to ethics. By contrast, they contend, an equally valid but distinctly different and more gender-sensitive approach asserts the centrality of interpersonal relationships, feelings, intuition, flexibility, contingency, and above all caring for others.

Carol Gilligan, a psychologist, is famous for her groundbreaking study of the moral judgment of women. She found among women of many ages a typical pattern of making moral judgments. This pattern consists of greater concern for the feelings and welfare of others and a more flexible approach to judgments that weighed contingencies. This contrasts with what she found to be the typical pattern of men, a rigid insistence on impersonal rules of justice that had to be indifferent to feeling and contingency. Though Gilligan's work is not without its critics, for many people it is highly significant for defining a distinct and important difference between the moral judgment of women and men and for asserting an entirely new approach to ethics. One of the most striking features of this approach is its opposition to the prevailing traditional view of ethics, which, assuming its masculinist bias, inappropriately denigrates the views of women while unfairly privileging the views of men.

UTILITARIANISM

Technical communication is a field that emphasizes usefulness as a key principle. It is fitting, then, that we examine the class of ethical theories that emphasizes usefulness. These are the utilitarian theories, such as that of John Stuart Mill, which weigh ethical judgments on the basis of accomplishing the greatest useful goodness for the greatest number of people. This amounts to a sort of technical or technological approach to ethics. The most prominent proponents of utilitarian ethics have been John Stuart Mill and Jeremy Bentham originally and Richard Brandt in recent years. For the sake of simplicity in our brief discussion here, we will treat utilitarian ethics as one.

Utilitarian theories not coincidentally rose to prominence at the same time as the rise of science and modern technological industrialism. These are movements that are particularly sensitive to the masses rather than to individuals. As a contemporary parallel to this view, we might think of the notion of the consumer in our own society. These are individual people of course, but as an aggregate people are of interest only as a means of consumption, not for their intrinsic personhood. Furthermore, they are important because of their numbers. Advertisers aim to sell goods, and the more, the better. So they advertise toward the greatest number of potential consumers. In the process, of course, they neglect the interests of individuals whose interests and needs might differ considerably from those of the majority.

A second connection between utilitarianism and the rise of science and industrial technology is that utilitarianism allows a quantitative calculation of what to do ethically. This avoids the messy indeterminacy of, say, Plato's trying to define clearly and correctly the nature of goodness or wisdom or of Kant's trying to determine how the categorical imperative might apply to a particular ethical dilemma. Instead, we can, presumably, decide fairly readily on an objective, quantitative indicator of utilitarian goodness, measure it for the appropriate number of people, compare it to measures of ill effects for the remaining people, plug it all into an algorithm, and calculate the solution. This presumably yields the best course of action in that situation. It is highly scientific or at least technical, allowing what amounts to a calculus of ethics.

In practical terms in our own times, consider an example. The Federal Aviation Administration many years ago made a judgment about whether to mandate the installation of fire detection and suppression technologies in commercial aircraft, including their cargo compartments. Recall that the ValuJet crash in the Florida Everglades in May 1996 involved a severe fire in the cargo hold that went undetected for some time and that could not be controlled. The FAA determined that, had fire suppression devices been installed, they might have allowed the plane to make an emergency landing.

The FAA many years before the crash had determined that fire suppression devices would undoubtedly save some lives and property over the years. The great cost of installing this equipment in comparison to the small numbers of lives, however, in the FAA's eyes, did not warrant mandating such installation. The greater good, from their point of view, was served by not mandating these changes, however. The low benefits for all outweighed the extremely high cost to only a few.

The explosion of the Boeing 747, TWA Flight 800, over Long Island Sound in July 1996 provides another illustration of both the common usage of a utilitarian approach in technological matters and the ethical difficulties of such an approach. The FAA's methodology of using cost-benefit analyses to decide whether to mandate that fuel vapor control or purging systems be installed in commercial aircraft has been criticized lately as being unethical. This cost-benefit analysis amounts to reducing the worth of a human life to dollars and cents. And of course, in the background, lies the additional question of whose dollars and cents are at stake.

Similar calculations occur throughout our modern society, such as in the Food and Drug Administration. Any new drug coming up for FDA approval for mass distribution must demonstrably prove itself to be safe and effective. It must show that it has a significant effect in easing some disorder and that has no serious side effects. Practically any new drug, we all know, however, has its side effects, some of which might occur frequently and some of which might be extremely serious. In its utilitarian calculus, the FDA must decide that the benefits outweigh costs and define the population to which that judgment applies. Some drugs are approved for general use to everyone except children, for example, or those taking certain other medications. Still, a weighing of competing consequences takes place, and the good must outweigh the bad.

In recent years, however, the FDA has been criticized for applying these rules too rigidly. Especially in the case of experimental drugs for the treatment of AIDS, critics contend that rules concerning possible negative side-effects are irrelevant when one is facing imminent death. In this case, critics contend, the conventional calculus is inappropriate; instead, the potential benefit—life itself for a few people—even for a short time, far outweighs the cost of a few deleterious side-effects. More important, critics contend that an impersonal bureaucratic agency that does not itself suffer through AIDS is specifically not the proper entity to make such calculations and decisions. Instead, the people directly involved should decide for themselves what is best for them.

Although utilitarianism is easy to state in principles, carrying it out in reasonable practice is another matter. Nevertheless, sometimes the utilitarian approach, despite its impersonalness and other difficulties, seems to be the only reasonable ethical approach to take. Medical ethics provides a good example of the validity of this approach in some situations. In seminars on applied and professional ethics such as are commonly offered in medical schools, students examine how they make ethical decisions. Typically they face a dilemma such as having to distribute a woefully inadequate number of donated organs to a large number of desperately needy potential recipients. The purpose of such exercises is twofold, to show us how complex, unclear, and painfully difficult real ethical dilemmas can be and to show us how to reveal to ourselves the tacit values behind our decisions.

Here is an abbreviated example of such a case. Suppose you had three human livers recently donated and a list of ten potential recipients. The circumstances force you to make a decision among the ten and soon. Three people will benefit and live long; seven will not and will die soon. At first you are given only a brief, fairly objective description of the ten potential recipients. This includes basic information about age, sex, current health, complicating medical conditions, marital status, and number of children. You have to rank your three selections and explain your choices. Presumably your selections reflect your true value system, what you really value above other factors rather than what you think you should value. Often in such exercises, young age and good general health are major factors. Through small group discussions about their choices, everyone sees what everyone else values when push comes to shove.

The second phase of the exercise provides additional information to complicate the decisions and to make them more realistic. One candidate, who is otherwise in very good health, might suffer from frequent bouts of severe depression and need to undergo extensive psychological and pharmacological treatment and might have threatened suicide several times. Another might be a young married person in good health who is known to have experienced an abusive childhood and has been charged with physically abusing others children in connection with her work. Another candidate might be a man in late middle age with a history of alcohol abuse but who is a famous baseball star; a transplant for him might bring a good deal of publicity to the doctors and medical facility involved. Similar new, complicating information is offered for all the candidates.

These new factors concern social, economic, religious, and political characteristics of the candidates, but they also, less obviously but no less importantly, concern the values of the people making the decisions. Many of these factors are often vague and uncertain, such as the realistic likelihood that a prison inmate will actually succeed in turning his life around. Do you judge someone solely on past behavior, or do you admit the possibility of change that is claimed but not yet demonstrated? Questions such as this also cause you to reflect on yourself—would you want to be judged by the same criteria that you are applying to others, or would you expect different treatment for yourself? As before, your revisions to your previous ranking reveal the values you hold most important.

Such exercises, now fairly standard in medical ethics courses, serve above all to clarify one's values so as to open them up to examination and possible alteration. In this way we are forced to confront how we really make our decisions rather than how we might wish to think of ourselves. We might alter a few of our values, or we might not. We might feel stronger in our convictions about our values. Often, even if no values are changed, we can leave with a clearer sense of why we hold them so important and with a new sense of confidence in what exactly we value.

FEMINIST AND CARE PERSPECTIVES

In this section we review feminist perspectives on ethics relating to technology and science and an ethic of care proposed by some feminist theorists.

The twentieth century has experienced widespread intellectual developments that have challenged many aspects of traditional wisdom. In the movement known as postmodernism, traditional authorities and knowledge long taken for granted have been challenged, while new perspectives are offered in their place. Modernism can be represented for our purposes by the traditions of rationalism, individualism, logic, analytic thinking, the advancement of science and technical knowledge, the view that knowledge is single and positive, the self as primal, and the idea of progress. These attitudes, assumptions, and values collectively have given us modern technological and industrial society, for better or worse. These notions have historically also been accompanied by unsavory companion notions such as the will to power, the desire to control, the subjugation of nature, and the increased capacity for domination and conquest.

Postmoderism seeks to replace modernist attitudes and values and to rectify their negative effects. It challenges authority, especially traditional authority figures such as religion. It also challenges what has been taken for granted, such as the desirability of hierarchical organization patterns or of scientific knowledge over other forms of knowledge. Knowledge from a postmodern perspective is complex, involving multiple forms, and is socially conditional rather than absolute. Indeed, the social and cultural context is much more important than the individual, which itself is socially constructed. Searching for the absolute foundations of knowledge is futile because of this social constructedness. Therefore, rather than to try to develop more and more objective, impersonal knowledge, it is more desirable from a postmodernist perspective to learn more about our societies and cultures and to open them up to critical examination. This criticism of traditional views applies to ethics, too, of course.

Although feminism is not intrinsically linked to postmodernism, a good deal of resonance exists between them. Even though there is no singular feminist theory or position on ethics, a good deal of resonance exists between some postmodernist thinking and some feminist thinking. Several important new perspectives on ethics have arisen from feminist critical thinking and from postmodernism.

Feminist Perspectives on Science as a Value System

Feminist critical thinking has expanded the meaning of ethics in important new ways. Though male critics of science and technology as value systems have also arisen in the twentieth century, most notably Jacques Ellul and Langdon Winner, feminist criticism is perhaps more unsettling because of the damaging discriminations it reveals. Feminist critics point out that both science and technology constitute value systems in themselves. They also reveal that these values, or at least the way these values have been carried out in practice, are fundamentally gendered. They have shown that the root values entailed in the scientific method, for example, have worked to the advantage of men and to the disadvantage of women and have privileged the work of men but disprivileged the work of women.

Sandra Harding and a host of other feminist critics of science including Ruth Bleier and Evelyn Fox Keller have examined the importance of dispassion in scientific inquiry, which all too often amounts to a total rejection of emotions altogether. Not coincidentally, historically women have been thought to be innately more emotional and more valuing of emotions than men. With science excluding emotions from its investigations, it is no wonder, critics such as Keller explain, that men have been drawn to science and women discouraged. They have critiqued, too, the analytical method of scientific investigations, separating things into parts to understand the whole better. Analysis, they contend, is basically a masculinist way of arriving at knowledge. Though analysis can have its merits, many feminist critics hold that analysis masks if not denies the fundamental importance of the organic whole, of how parts operate together as a unit, and how the organism as a whole exists.

Scientific investigations often focus on isolated elements rather than complex interrelations, critics add. They emphasize the damaging effects of an excessive preoccupation with the individual, such as is found in much of psychology, rather than with complex social and cultural interrelations. Many feminist critics of science contend that this impulse to separate and isolate and thereby to minimize if not nullify relationships, is characteristically masculine. A feminist sort of science, some contend, would emphasize the whole organism and the entire interrelational social complex in which organisms fully live.

Dispassionate logic with its rigidly binding conclusions is the method of reasoning used by scientists, in addition. Logic, some feminists critics contend, as a basis for thinking is more characteristic of men than of women. Thus, again, men feel comfortable thinking logically and doing science and so thrive in this environment, whereas women feel uncomfortable and do not thrive. This accounts for the de facto discrimination found in most of the scientific and technical fields, which seems almost necessitated by its fundamental principles and attitudes. (See for instance Jo Allen's "Gender Issues in Technical Communication Studies" and Sam Dragga's "Women in the Profession of Technical Writing.") Mary Field Belenky's *Women's Ways of Knowing* is perhaps the most important scholarly exploration of gender differences in what counts as knowledge and how it is developed.

Some feminist critics of science further point out that the underlying purpose of scientific and technological investigations is usually to predict and therefore to control. Usually what is controlled is nature, which, not coincidentally, is usually characterized as female Mother Nature, whereas those who are doing the controlling are scientists, usually males. Annette Kolodny, in *The Lay of the Land*, for instance, traces the masculinist impulse underlying the explosive technological and industrial expansion in most of American history, which aimed at subduing and controlling nature. It was pointedly insensitive to environmental destruction, what is often called the rape of the landscape.

These feminist criticisms are not without their own critics, however, from other women or feminists. These critics point out the essentialism and biological determinism assumed when one says that something is characteristic of one gender or another. Notice, for example, that Nel Noddings' ethics of care, discussed later, presupposes essential differences between the sexes that in turn determine behavior differences. They point out that the more accepted and less discriminatory view is to take gender-associated behaviors as largely, if not entirely, socially constructed. Evelyn Fox Keller's essay "The Gender/Science System: or, Is Sex to Gender as Nature is to Science?" is particularly enlightening in this area. Keller's famous book on the biologist Barbara McClintock, *A Feeling for the Organism*, which paved the way to McClintock's belatedly being awarded the Nobel Prize, illustrating feminist criticisms of science, is a very real example.

Critics also point out some potential difficulties in characterizing logic as a characteristically male way of thinking. To do so seems to assume that women are inclined to be illogical, that is, unreasonable or incapable of reasoning clearly and correctly. This implicitly affirms a stereotype under which women have already suffered too long. And the other side of the coin can have a similar effect. To assume that emotionality and sympathy are characteristically female ways of relating to others (e.g., Noddings and Gilligan) can also implicitly affirm the damaging stereotype that women are feeling but not thinking people and are best suited to nurturant roles but not scientific roles. At the same time, this can implicitly affirm a damaging stereotype of men, that they are unfeeling and uncaring.

These glimpses show the diverse positions among feminist thinkers on ethics in relation to science and technology. This diversity should not be surprising, for flexible variety rather than rigid singularity is a basic principle of a good deal of feminist criticism. Not only is complex variety encouraged, but there is also found among some feminist critics a positive affirmation of open-endedness and continuing dialectical interchange such that dialogue is never closed off and issues are never settled once and for all. *Conflicts in Feminism*, by Marianne Hirsch and Evelyn Fox Keller, for example, celebrates diversity of opinion and openness to alternative views.

The implications of feminist criticism for technical communication are profound. We need of course to be careful to avoid sex-related terminology in order to avoid bias, stereotyping, and discrimination. But we also need to be conscious of all of our terminology and its unavoidable value-ladenness, which is often inapparent or denied. Mary Daly's *Gyn/Ecology: The Metaethics of Radical Feminism* is especially clear in explaining the many powerful ways that implicit values inform our language and thinking at the most fundamental levels. As Kenneth Burke has pointed out, our language operates as terministic screens, allowing us to see things differently depending on the screen. We can, however, never *not* use some sort of terministic screen because selection and reflection are inherent to language as symbols. All we can do, and all we should do, is be conscientiously aware of the values communicated by our language.

Feminist critical thinking can be applied to technical communication in many ways. We can readily see the gendered aspect of terminology such as "joystick" and "cockpit" in aviation, of "male" and "female" couplings in electrical and hydraulic devices, and of computer codes as "commands." Less obvious but no less value laden are the competitive and militaristic, and therefore supposedly male, language and images used to sell practically anything technological. We are told, for example, that a new piece of equipment is on "the cutting edge" and so will give us a "competitive advantage" that will lead to market "dominance." This frame of mind assumes a Darwinian sort of industrial environment in which only the fittest survive, and these presumably will be the strongest, quickest, and least caring for others. Feminist critics, as well as other ethical critics, point out the need to rethink our society and economy to create a win–win environment in which all can thrive. Hierarchical organization in corporations, which is associated with a masculine view of society, for example, is giving way to more horizontally stratified, decentralized, and egalitarian organizational structures.

Many articles on technical communication from a feminist ethical perspective have been published in recent years. We should recognize that the influence of feminism generally and its revelations about the gendered nature of not only social policies but also values systems and language use have given rise to studies of gender issues, too. Probably the most famous of these articles is Mary Lay's "Feminist Theory and the Redefinition of Technical Communication." Lay reviews six key characteristics of feminist theory generally, then explains how they should be applied to redefine the field of technical communication. Another important article by Lay, "Interpersonal Conflict in Collaborative Writing," describes some of the difficulties stemming from differences in perception between men and women about interpersonal conflict. This article also mentions the issue of essentialism and biological determinism versus social constructionism as the source of our behavior. Whereas men typically value interpersonal conflict, women tend to avoid disagreement. Because typically male attitudes toward interpersonal conflict historically have predominated in complex organizations, women's attitudes and expectations have often been neglected. This fosters discrimination and results in the organization's failing to benefit from the potential but unrecognized contribution that women could make.

Another important article is Jo Allen's "Gender Issues in Technical Communication Studies: An Overview of the Implications for the Profession, Research, and Pedagogy." Allen focuses on the systematic discrimination against women in the profession, which is quite similar to that in other professions. She traces the bias against women's typical ways of knowing and interacting, which stress harmony and conciliation rather than confrontation or decisive action. Allen also shows how women encounter discrimination in lower earnings and restricted advancement. This result from women's work not being valued properly, simply because it is done by women rather than men. Value is the point here, both philosophically and monetarily.

Feminist critiques are also applied to our field by Lee Brasseur's in "Contesting the Objectivist Paradigm: Gender Issues in the Technical and Professional Communication Curriculum." Brasseur challenges common notions about scientific and authorial objectivity and disinterestedness. The aim is to enact the suggestions of other feminist critics "to replace a discourse model which emphasizes expediency at the expense of social and cultural awareness with one that speaks to multiple positions and moves toward a new paradigm for 'objective' discourse" (115). Brasseur accomplishes this pedagogically through a series of readings and a program of ethnographic research by students. This research is deliberately opened up to critical appraisal in class in order to reveal the value-ladenness of a supposedly objective frame of mind, to reveal the invidious effects of objectivist research on subordinate social groups, and to reveal the validity of alternative, nonobjective methodologies. The critical examination of and flexible choosing among value systems lies at the heart of Brasseur's pedagogy.

In a similar movement from feminist theory to applications, Beverly Sauer reports on the mining industry in "Sense and Sensibility in Technical Documentation: How Feminist Interpretation Strategies Can Save Lives in the Nation's Mines" and in other articles. Sauer offers both a feminist critique of, and a feminist prescription to correct, the powerful yet unacknowledged gender assumptions in traditional technical communication in the mining industry. Sauer is particularly interested in how the voices of women have been systematically devalued and silenced in the investigation of mining disasters. The voices of men, on the other hand, have been systematically valued and listened to because of a fundamental gender bias in the technical discourse of the mining industry. This bias stems from value assumptions to the effect that males are characteristically objective, rational, and factual—and therefore sensible. The other side of the coin is that females are assumed to be emotional, impressionistic, anecdotal—and therefore less meaningful for technical discourse. Sauer concludes: "A feminist analysis demands that technical writers acknowledge the silent power structures that govern public discourse, not because we are interested in theoretical constructs about language but because those power structures affect the fabric of technology on which we all depend" (523). Language reflects values that in turn reflect power.

To summarize, feminism amounts to a system of values, an ethic. A feminist perspective on ethics in technical communication brings to our attention many important factors that would escape our notice through conventional ethical approaches. It reveals the bias against women and women's work in a variety of ways in technical communication, for example, and it questions the importance of objectivity as a preeminent value while it challenges the assumption that objectivity is even attainable.

In other ways, however, a feminist approach lends additional support from a different quarter to some conventional ethical approaches. The caring concern of some feminists echoes Kant's call for the treatment of any people, whether men or women, always as ends in themselves, never as means to others' ends. It echoes, too, ethical critics such as Ellul and Winner arguing against the powerful but tacit assumptions underlying the ascendancy of technology as a basic force undergirding our culture.

Most important, it requires that we open up to critical examination the very roots of what we take ethics to be. It questions the universality and gender-indifference traditionally supposed of ethics, and it questions whether ethics is not historically gendered in the sense of valuing men over women. It questions the principle of absolute justice usually assumed to be fundamental to ethics while positing care or other concepts as equally valid alternatives to justice, and it questions the role of reasoning alone in guiding ethical judgment. And it questions whether ethics can be subsumed under a definite system of principles once and for all or, on the other hand, whether ethics is never one thing but many things that are constantly evolving as our conversation shapes our ethics. These are questions of the most fundamental sort, and, because ethics is always about questioning, weighing, and critiquing ourselves, feminist criticism is of great service to ethical studies in the most fundamental way.

Ethics of Care

There are many different perspectives on feminist ethics. Among the more prominent names in the movement to try to articulate a feminist view on ethics—or at least an ethic sensitive to gender differences—are Carol Gilligan, Nel Noddings, Mary Jeanne Larrabee, Sandra Harding, Evelyn Fox Keller, Virginia Held, and Rosemarie Tong. One of these perspectives is an ethics of care. (Some theorists categorize ethics of care that focus on women's unique way of knowing and of making moral judgments as "feminine" or "femininist" ethics rather than "feminist.") We can glimpse here only one of the many contributions on ethics of care.

The ethic of care of Nel Noddings presented here does not represent all or even most feminist approaches to ethics. Though some feminists support the idea of an ethic of care, many others find compelling reasons to reject it as a specifically feminist sort of ethics, preferring to refer to it as a feminine or even "femininist" ethic. Noddings is presented here only as an illustration of the range of new possibilities in the field of ethics afforded by the critical thinking of postmodernism and feminism. Many of these possibilities directly challenge some of the assumptions underlying traditional approaches to ethics and depart markedly from them. As we will see in the next section, some feminist thinkers object to the assumptions in Noddings' ethic of care. There are also many legitimate forms of an ethic of care. Our focus here is only on the version represented by Noddings informed by the work of Carol Gilligan and others.

Carol Gilligan, mentioned earlier in relation to Kant, conducted groundbreaking research in the role of gender in ethics in her book *In A Different Voice*. Though part of her work focuses on the role of language in communication between and about men and women, the main thrust of her work is to characterize the difference in moral thinking between men and women throughout the human life span. Women typically feel that relationships are of paramount importance, according to Gilligan. Both the relationship, marked by a caring concern, and the other person are generally valued more in women's moral judgment and ethical decision making than in men's. Men, on the other hand, Gilligan says, base ethical decisions on justice and are more inclined to think of ethics impersonally, as a matter of abstract principle such as Plato, Aristotle, or Kant did. Ethical principles exist apart from and prior to the person and therefore are to be taken as more important than the particular person. Simply put, women generally emphasize caring concern, relationship, and the flexible application of values depending on the particular person and circumstance in rendering their ethical judgments, whereas men generally emphasize justice through the inflexible application of abstract principles regardless of the person or the relationship.

This difference in moral thinking is generally unrecognized, Gilligan explains. Traditionally, the typically male way of thinking about ethics has come to be taken for granted as applying equally to men and to women and to humankind generically. The difficulty with this overgeneralization is that it devalues or just simply disregards women's judgments. This silences women's voices and excludes their participation in what are intrinsically important matters, (e.g., ethical judgments). It also implies that the way women typically make ethical judgments is flawed, incorrect, and invalid, thereby reinforcing their silencing and exclusion.

Gilligan's findings greatly complicate any discussion of ethics. Rather than trying to decide on one system of ethical principles over another, Kant over Aristotle, say, or of traditional ethics as a group, one must recognize that such systems are inherently invalid for fully half of humanity, which operates in another ethical realm completely. As a consequence, not only would we have to revise the way we think about ethics, we would even have to rethink our language about ethics.

Nel Noddings in *Caring: A Feminine Approach to Ethics and Moral Education* resonates with Gilligan in many ways, particularly in rejecting ethics based on justice and impersonal, abstract principles. She is careful to represent her position as different from but not necessarily oppositional to traditional masculine views on ethics, however. She positions caring at the center of ethics and emphasizes the principles of receptivity, relatedness, and responsiveness. Ethically responsible caring does not mean, for instance, a subordination of one's own welfare and happiness to another. Radically selfless caring, Noddings explains, which has traditionally been expected of mothers and women in general but not of men, actually fails to be responsible to oneself. Noddings instead advocates an interdependent relationship of caring among equals that is mutually satisfying to all parties.

Mary Jeanne Larrabee, editor of *An Ethic of Care*, critically examines the idea of caring as a basis for ethics and the idea of whether there can or should be an ethic specific to either feminists or women. The idea of an ethic associated specifically with women is problematic, according to some feminist thinkers. One of the dangers of trying to develop a system of ethics specific to women is that it can seem to confirm traditional stereotypes while distancing itself from what is commonly thought of as legitimate ethics. Essays in her book span a range of positions and represent effectively the complexity of these issues.

Peta Bowden reports that many feminist theorists object to various ethics of care whether founded on mothering, friendship, or nursing. Their objections fall roughly into three categories. One, they "fail to take into account the oppressive conditions in which many women's practice of caring occur," which are usually characterized by subordination and dependency (8). Two, they reflect an impoverished situation in which care is not reciprocated and in which only limited sorts of relations are possible. Three, they assume a biological determinism that constrains the expectations of women's behavior and denigrates any divergence from such expectations.

Interestingly, Bowden notes that the dichotomy between caring and justice seems to correspond to the dichotomy between the private and the public realms but that neither is absolutely necessary. Civic societies do not have to be impersonal and uncaring, that is, though historically most have been. Bowden, building on the work of Martha Minow, suggests instead that the possibility of an alternative ethic of care grounded in citizenship that embraces caring and egalitarian principles should be explored.

Virginia Held in *Justice and Care: Essential Readings in Feminist Ethics* presents a range of women thinkers on the relation between ethics of justice and ethics of care. Important statements on both sides regarding the validity of an ethics of care are presented by several theorists. Others seek to integrate ethics of justice and care while mitigating the essentialism implicit in most distinctions between them.

Rosemarie Tong's *Feminine and Feminist Ethics* offers a comprehensive examination of the differences and commonalities between feminine and feminist ethical theories and theorists. (Most ethics of care would be categorized as feminine theorists in Tong's scheme.) Tong also compares and contrasts these theories to traditional ethical theories such as utilitarianism and deontology.

OTHER VIEWS

In this section three ethical theories are described briefly: those of Confucianism, Emmanuel Levinas, and Bernard Gert. They are presented to suggest the diversity of ethical perspectives across time and cultures, the full range of which is beyond the scope of this book. The reader is urged to explore more fully the theories presented here and others through further study and discussion.

Confucian Ethics

To give us a broader understanding of what ethics means, it would be useful to consider an ethical perspective from a distinctly different culture. This will help us to see more objectively the ethics of our own culture. It will also leave us with a clearer sense of responsibility for actively choosing our own ethics.

The description of Confucianism and Confucian ethics that follows presents only the most basic and enduring features of the Confucian moral code over the twenty-five centuries of its existence, though there have been numerous variants of Confucian thought over the years (derived largely from Dawson, Hansen, and Wei-ming). Confucian ethics is a part of Confucianism, a philosophy that has profoundly influenced much of the non-Western world. At one time or another it has influenced practically all of eastern Asia, including China, Japan, Korea, and Vietnam, with its vast populations and sophisticated cultures, an influence that continues to this day, though now in somewhat muted form. The second half of the twentieth century has brought fundamental changes to the Asian world, however, especially to China. The most significant political and social development in China has been the introduction of Communism, which gave birth to the People's Republic of China. This new social, economic, and political order, the result of a revolution, has insisted on the renunciation of the old order, including the Confucianism underlying that order.

In recent years, though, there seems to be developing an accommodation of sorts among Communist leaders to some of the traditional tenets of Confucianism, in part because of their intrinsic merit but also because of the simple endurance of their influence throughout Chinese society. Certainly the Confucian emphasis on collective entities (whether as family, state, or nation), for instance, resonates strongly with the collectivism that is the essence of Communism. Interestingly, this effort by the Communist regime to reshape past notions in light of present circumstances is much like one of the basic principles of Confucianism itself.

Confucianism and its ethics are grounded in immediate realities rather than in immutable, timeless absolutes. Confucius himself discouraged speculating too much about the spiritual realm because the immediate, real world is the stage upon which morality is truly played out. Confucian ethics defines human responsibilities as being constituted in relationships, not in the isolation of a radical individual. It also insists on the subordination of individual egos to time-honored obligations of social relations and to the needs of social harmony.

Confucian ethics asserts the fundamental importance of several key principles. The active practice of these principles by a person constitutes morality and ethical conduct. These ethical principles have a concrete realness that differs considerably from Platonic transcendent abstractions. One's morality, for example, consists primarily of one's behavior toward others, through demonstrations of respect and deference and through the practice of conventional courtesies. Though the cultivation of proper attitudes behind these actions is important, attitude without action is meaningless. Furthermore, one's behavior in relation to real, immediate circumstances is valued more than adherence to abstract, absolute principles. This circumstantial context extends from the immediate present far into the past in the form of traditions and ancestral relations. Confucius himself, for instance, developed his philosophy to propagate the values of an ancient golden age of Chinese society. By defining one's self only in relationships in this way, Confucianism is collectivistic rather than individualistic.

Confucian humanism also differs from traditional European-American humanism in the way in which its ethical principles are arrived at. In the European-American tradition, we have Aristotle dissecting and investigating any given topic through abstract theorizing on the basis of first principles to yield a systematic organization of knowledge. We also have the determined rationalism of Kant, who grounded his inquiries on the bedrock of the rational mind's being able to figure out the nature of values and ethics. The result is a systematic argument developing toward a logically and ethically binding conclusion.

Confucianism, on the other hand, approaches the topic of ethical virtue differently. What we know about virtue, according to Confucianism, does not rest on a foundation of logical or analytic reasoning but on virtue itself as exemplified in people and episodes. In this way virtue and our language about it are understood to refer to definite, unambiguous, real entities, not to abstractions. If you want to understand, say, honesty, you study the story of a particular individual who exemplifies that virtue. This knowledge about ethical values is learned through the study of particular historical cases and through studying the body of learned interpretations of these cases from worthy scholars. The *Analects* of Confucius, for example, consist of numerous short passages, something between proverbs and parables, that provide terse statements on valued behaviors. The passages are so short and so descriptive rather than explanatory that numberless scholars have worked to interpret them properly. In this way Confucian ethical study can be characterized as hermeneutic (that is, involving interpretation rather than, say, analytical reasoning), an approach to knowledge that has only recently become important in mainstream European-American philosophy. Another difference is that ethics is constituted in the lives of particular persons—how and why they act in relationship—and so cannot be called knowledge in any conventional sense because it involves *doing*.

The most important Confucian principles are *ren*, *li*, and *yi*, which together constitute a general ethical virtuousness called *te*, which the good person must cultivate carefully. The cultivation of morality is of paramount importance because it is the basis for any and all authority and is the foundation of a happy, productive society. The emperor, for example, can have authority and is justified in expecting deferential respect from subordinate officials only because he first is a moral person. (One is reminded of Cicero and Quintilian in the Roman tradition.) Thus political authority is fundamentally a matter of morality rather than of power or legality.

One cultivates virtuousness by understanding and carrying out many activities that together compose the single "way," tao, of virtue. This way embodies our proper relationship to the universe, including heaven but more importantly the social realm and serves as our guide. It involves thoroughly understanding the nature of virtue by learning the ethical traditions and the interpretations of exemplary lives from worthy teachers. It also involves understanding the obligations and duties entailed in various relationships and executing these piously, first and foremost through the "filial piety" that a good son shows toward his parents.

These duties and obligations are ethically carried out according to principles of propriety, *li*, in the form of traditional rituals. The dimension of tradition is crucial here in acknowledging that one's place in the world is defined only through relationships and only by affirming one's connections with venerable traditions transmitted over the ages. These obligations are binding for any given relationship, though the strenuousness of the obligations decreases progressively as one moves farther from the central family relationship. The result is a social hierarchy that is characterized by gradations of loyalty or obligation. Thus one is bound by relationships to everyone in society but not all to the same degree or in the same way.

The careful practice of rituals of piety is not done for the sake of ritual itself, however, but for the conscientious enactment of a more fundamental sense of rightness or appropriateness, *yi*. This virtue can also be understood as a sense of justice, though a justice defined in terms of one's social station. This justice operates to maintain the social hierarchy and traditional relationships, which are assumed to be intrinsically just. One's ethical burden lies in conscientiously fulfilling the duties and obligations associated with the particular web of relationships in which one finds oneself (and specifically not in undermining these relationships or challenging these obligations).

At a still more fundamental level lies the virtue of humaneness or humanity, *ren*. This humaneness is one's sense of oneself as a human being and as related to all other people through our common humanity. Thus Confucius can say that ren is the love of all people. Without this fundamental awareness and attitude of love, genuinely virtuous action is impossible.

At the same time this humaneness involves a respect for others because they share humanity with you and with everyone else. It also involves treating others as one would like to be treated by others, the Confucian analogue of the Western Golden Rule expressed negatively: "What you do not want for yourself, do not do unto others" (*Analects* 15:23). In addition, it involves the deliberate cultivation of humaneness, a personal study of what it means to be fully human. This requires the rigorous, lifelong study of texts, history, and all aspects of culture and social relations. Only through studying the example of others can we learn how to become our human selves as fully as possible, though in being fully human we are really only expressing what already lies within us. (One is again reminded of Quintilian.)

The paradigm illustrating how these principles are to be carried out is the relation of the son to his parents. Confucianism holds males as primarily responsible for most ethical activities, with females having responsibilities derived from relationships with males, as in showing ritual respect for her in-laws or in bearing and raising children. The son is responsible for caring for his parents as they live, for burying them with proper rituals, and for ritually revering their (and his) ancestors. This ties the son through the generations as microcosm to the rest of creation as macrocosm, reflecting his recognition of the common humanity of us all. It shows a sense of justice or appropriateness as he cares for those who cared for him. At the most fundamental and socially important level, it shows his basic virtuousness, for if he cannot show proper respect and care for his own family, how can he show proper respect and care for the rest of society in any other relationships?

This sense of common humanity is not egalitarian, treating all people as identically equal, as it is in traditional European–American ethics, however. Instead Confucianism explains that a person finds himself or herself in a particular social situation that should be accepted and respected. Though we are all equally obliged to behave ethically, the way in which that obligation is carried out depends on the web of interrelationships one happens to live in. The parent-son relationship, for instance, has a reciprocal dimension by which the parents have corresponding duties and obligations to the son, but they are not equal in force or depth to the son's toward them. Thus the particulars of ethical behavior are determined principally by social context rather than by absolute, transcendent principles that are indifferent to the person or the social context, as in Kant's ethics, for instance. For similar reasons, the principle of justice often amounts to appropriateness, the course of one's ethical actions being guided by what is appropriate to the hierarchical social context. Thus from a Confucian ethical perspective, we should not expect everyone to behave the same to all people everywhere. One is not obliged in the same manner or in the same degree to an employer, say, as one is to one's immediate family or to one's close neighbors. Likewise honor and trust are contingent on social circumstances. In many modern European–American societies, by contrast, a sense of fairness that sees all people as interchangeably equal is taken for granted.

Although the particulars of any ethical obligation depend on the social circumstances, still the need to be ethical is fundamental. In government and throughout the rest of the social order, the leader should be a paragon whose ethical conduct garners support and obedience from those governed. The result, ideally, is a rule not by force or legal power but by moral example, yielding a nation of harmony and happiness.

Coupled with these ethical principles emphasizing relationships is a traditional system of social organization reflecting these principles. Traditional Chinese society is a structure of hierarchical relationships, each of the five major grades in the hierarchy having its own particular obligations and duties. The root of this hierarchy is the relationship of son to parents in which the son enacts the principle of filial piety and which serves as a paradigm for other, more distant relationships.

Each member of society has many duties and obligations defined by his or her relationships. A man, for example, would be engaged in a relationship with his father but might also have a son to which he must relate as father. Stemming from this root relationship are those between emperor and officials, husband and wife, elder and younger brothers, and friends. Included in this lower part of the hierarchy is the employee-employer. In recent years, however, employee-employer relationships have become strained and problematic due to the powerful opposing influences of communism and industrial capitalism.

The family metaphor resonates through all levels of society, the strength of the obligations involved weakening as one moves farther from the core relationship of the family. Indeed, as one of the later interpreters of Confucius' teachings put it, to treat friends or colleagues the same as one's parents is to demean and insult the parents. Thus one should not expect the same degree of adherence to virtues through the different social strata. Because the strata are not themselves equivalent in importance for one's innate humanity, to treat all relationships the same would violate justice, too.

How does all this relate to ethics in technical and professional communications? Admittedly many East Asian societies, especially China's, are in a state of ongoing change representing an intermingling of Confucian and other traditional ethics with the distinctly different ethics of communism and industrial capitalism. Nevertheless, it is safe to say that traditional Confucian ethics still play an important role in shaping relations and communications throughout Chinese society, including the technical and industrial realms.

Typical technical communicators in the United States dealing with Chinese industry can expect a clash of cultural values to some extent in their relations and discourse. This clash, though awkward, need not be a serious difficulty. Keep in mind that Confucianism traditionally places great value on the gracious and harmonious accommodation of venerable traditions to new, changed circumstances.

Tradition continues to be prized in the form of rituals of courtesy and deference and in the form of compelling obligations stemming from relationships and actions. By contrast, Americans, for example, typically value innovation, even iconoclasm, as do American corporations at times. More important, from a Confucian perspective the neglect or violation of obligations and proprieties carries serious consequences because they violate not just convention but ethics itself.

In the corporate context, traditional Chinese business relations were modeled on the paradigm of the parent-son relationship. The traditional relationship between employees and management therefore would appear to European-American sensibilities as paternalistic, authoritarian, and rigid. This relationship was based on the need for harmony, cooperation, and self-continuation as a valued tradition. Deviations from these expectations by an employee were considered very serious and were usually dealt with by socially shunning the violator. Thus in a traditional Confucian business situation, conflict or direct confrontation of management by an employee was highly unusual and very problematic. Contrast this to the contemporary American business context with its institutionalized affirmation of the confrontation of management by labor in the form of unions, for example, and with its laws protecting whistleblowers.

The ethical situation is much more complex than paternalism, however, because of the reciprocal duties and obligations involved. A business traditionally was seen as an extension of the parental model with comparable responsibilities to treat employees with respect and care. Furthermore, businesses were understood as integral parts of the entire social fabric, all aimed at working together cooperatively for the general good of society. Businesses were not ethically free to act arbitrarily or abusively toward their employees. Indeed, there is a clear strand of traditional Confucian thought affirming the need of subordinates to rectify or reprove their leaders and to remove them if necessary.

In relations with those outside the circle of employees—the general public, that is—businesses also have important obligations and responsibilities. Simply put, businesses are expected to enhance the common good, working cooperatively with government and compatibly with other businesses in a harmonious overall social structure that serves that good. The antagonism and suspicion between business and government that we often see in contemporary America, for instance, would not have been tolerated in traditional Confucian society. In addition, there is a long tradition of disdain of profits as the primary motive for actions. Not only were ethical principles elevated above the desire for profits, but profit was at times considered an outright threat to social well-being. Therefore, in traditional terms, management could not justify its activities on the grounds of enhancing profitability because the "bottom line" was not profits but morality and the cultivation of the common good. Thus the general business relationship that might at first appear to us as unethically authoritarian was kept from being arbitrary, abusive, or antisocial because of its underlying ethic.

Owing to the traditional social structure with its progressive gradations of loyalty and obligation, diluted as one moved farther from the central family relationship, some of our common expectations in the United States about ethics in professional dealings simply do not hold. From the Confucian perspective, it does not make sense to hold the same standards of loyalty, deference, honesty, integrity, or other virtues in our business dealings as we do in our closer relationships such as to friends or family. It would also be a violation of justice to do so.

To summarize, traditional Confucian ethics involves some of the same values as many European–American ethical perspectives: justice, honor, respect, humaneness. Donald Etz, for example, has shown how many selections from *The Analects* readily apply to ethics in technical communication in principles such as honesty, excellence, trustworthiness, deference, clarity of language, and acknowledging one's limitations. In other ways, though, a society built around Confucian ethics is quite different from one built around, say, Kantian ethics. The Confucian society is more concerned with preserving tradition and with performing proper ritual courtesies. Its social order, including the professional realm, is rigidly stratified. The acceptable social interactions associated with relationships within and between these strata is more formalized, too. And transgression of these social proprieties is more heavily sanctioned.

All these observations have to do with an idealized Confucian society. The ideal, though, has never existed historically. In addition, as the world moves toward a global economy and different cultures intermingle and strive to coexist, the traditional Confucian ethical ideal cannot be taken as representative of contemporary Chinese culture. On the whole, however, as this example suggests, a good deal of commonality among the world's ethical systems allows any ethically sensitive person to circulate comfortably almost anywhere. Respect, honesty, good will, and the gracious accommodation of differences are valued by almost all cultures, as is the person who cultivates these values in oneself.

Levinas

Among postmodernist ethicists of recent years, one of the most highly regarded but also one of the most abstruse is Emmanuel Levinas. Levinas seeks the root of ethics and finds it not in abstract principles or systems of values derived from rational analysis but in the particularity and uniqueness of our encounters with other people, which he refers to generically as "the other."

It is difficult to outline Levinas's ethical perspective because it challenges the powers of language to articulate feelings and thoughts. Levinas contends that ethics is not an abstract or metaphysical system of principles, nor a rationally understood sense of duty, nor a computational weighing of costs and benefits, nor a feeling of kindness toward others. It is, rather, about our human nature in relation with others. Ethics, Levinas says, result from our awareness of the other. The other refers to our sense, when we communicate with another person, that something utterly different from ourselves is confronting us, making us aware that we are not alone in the universe and certainly not the center of it.

The other makes us aware that some other thinking and feeling human exists, whose wants, values, feelings, thoughts, and responses are radically unknown to us and can never be fully anticipated. To determine how we should behave in relation to that person, we must understand what that person needs and wants from the relationship with us. This can be known only through communication in a give-and-take interchange that recognizes in the other person a morally equal yet unknown factor. We cannot make judgments or inferences about what is ethical conduct vis-à-vis that person unless we first communicate reciprocally with her or him. For this reason, rationalism as a basis for ethics is rejected by Levinas. Ethical principles or laws, as universal generalizations and abstractions, are also rejected.

To illustrate how Levinas' philosophy of ethics might be applied in practice, let us consider an example (for details see Dombrowski's "Can Ethics Be Technologized?"). The Vietnam Memorial in Washington, D.C., is famous as one of the most powerfully moving architectural productions of our times and one that makes a moral statement. As a structure, it is simple—a long, low wall of polished marble sunken in the grass like a grave. On its face is engraved the name of each person who died in the Vietnam war. The monument is experienced by walking across the long face of it, right against the names, which cannot be evaded.

Anyone who has seen it knows its deep emotional and ethical impact, which is to make the observer personally involved with and responsible to the people who died in that war. It does this by forcing the observer to recognize the unique personhood of each and every one of those individuals through their names. The passing observer, furthermore, can see the living reflection of her or his own face superimposed on the names of the dead—the contrast is striking. In this way the observer literally as well as morally *faces* herself or himself as a result of *facing* the names of each of the dead. As a result the observer feels both attached to and ethically responsible to that *other* person and for that death.

The uniqueness, both of your own personhood and that of the dead, somehow transcends whatever feelings and attitudes you might have brought to your visit to the monument. You are compelled to abandon whatever general attitudes or abstract ethical principles you might have been carrying as you face the uniqueness of these real other persons.

For Levinas, "the other" is ethically even more important than "I." Traditional ethical theories begin with "I" as the root of ethics. The personal subject, "I," thinks things through a dilemma, makes an ethical judgment, and wills to specific action. Levinas, however, rejects this personal subjectivity, instead insisting on the primacy of our experiencing of the other. This makes us aware of someone different from us—including all our principles and preconceptions. As a result, the differentness of the other makes us aware of being responsible—and responsive—to an entity equal to yet other than ourselves. Only the other can make us aware of the impact of our behavior on the other so that we can adjust our behavior responsibly.

Gert

A prominent ethicist of the late twentieth century, Bernard Gert has focused on the topic of morality and explored what it means, how it is known, how it relates individuals to society, and how it is carried out practically. His perspective has resonances but also differences with the perspectives we have been examining. It relates morality to rational thinking, to a universal audience, and to traditional moral principles while carefully distinguishing it from emotions such as caring, from religious duty, from personal authenticity, and from several other bases of morality proposed throughout history. Its basic definition of morality, equivalent to ethics, is this:

> Morality is a public system applying to all rational persons governing behavior which affects others and which has the minimization of evil as its end, and which includes what are commonly known as the moral rules as its core (6).

Morality for Gert involves action (rather than feelings), social relations with others (rather than absolute relations with god or to abstract principles), applied impartially to all including oneself, and the avoidance of evil rather than the pursuit of good. These elements are represented in common moral rules whose rightness is recognized by all rational people (rational not in the sense of coldly logical but rather as just not irrational). Gert's derivation of what constitutes morality is rigorous and spans the entire ethical heritage from Plato, through the Golden Rule and the Ten Commandments, Kant, Mill, and Moore, to contemporary ethical theories based on authenticity or caring concern. His language throughout is accessible to the layperson (using terms such as good, evil, virtue, vice, ideals, and judgment). Gert emphasizes the avoidance of evil as being more definite and decidable rather than the pursuit of good, which he feels is too indefinite and can take many different forms that can exemplify different degrees of goodness. Most people can readily agree when pain is being caused to another, though it would be difficult for them to say when another is pursuing or feeling pleasure or whether another sort of pleasure might be more morally desirable.

The five primary moral rules all infringe on another's life and person and have roughly to do with what most people would call violence or violation:

1. Don't kill.
2. Don't cause pain.
3. Don't disable.
4. Don't deprive of freedom.
5. Don't deprive of pleasure.

The second five have roughly to do with honesty, fairness, keeping promises, and being true to your words through your actions:

6. Don't lie.
7. Keep your promises.
8. Don't cheat.
9. Don't commit adultery.
10. Don't steal. (284).

Though carrying out these rules consistently would make one appear to be virtuous, Gert is careful to avoid reducing morality to personal character, which would make good or bad persons the explanation for good or bad behavior.

Gert offers a short list of eight questions to be used in determining the morally relevant features of a given situation. Though the questions are few and short, the answers in practice would be long and complex, of course. The questions are these:

1. What moral rules are being violated?
2. What evils are being (a) avoided? (b) prevented? (c) caused?
3. What are the relevant desires of the people affected by the violation?
4. What are the relevant rational beliefs of the people affected by the violation?
5. Does one have a duty to violate moral rules with regard to the person, and is one in a unique position in this regard?
6. What goods are being promoted?
7. Is an unjustified or weakly justified violation of a moral rule being prevented?
8. Is an unjustified or weakly justified violation of a moral rule being punished? (285).

To summarize, Gert provides a detailed, rational understanding of how and why to behave morally in today's society including the professional and technical realms. It utilizes many everyday notions and terms that we can all readily relate to and that stem from the very ethical theories we have been studying. At the same time, he integrates them into a single coherent system that is relevant to today's world. These are applied in a number of detailed case studies. Though most of these cases have to do with medical ethics, many are readily translatable to a technical and professional communication context. The result, Gert explains, is "a system that people can actually use in dealing with real moral problems" (282).

CONCLUSION

In this chapter we reviewed some of the most notable theories of the European–American ethical tradition. These range from the virtue ethics of Aristotle, through the rationally derived ethics of duty of Kant and the practical ethics of consequences of utilitarianism, to the contemporary ethics of care challenging traditional notions of authority, justice, and ethical foundations. These perspectives offer the concepts and vocabulary that correspond to our everyday notion and yet taken together help us to develop our own particular ethical perspectives.

Other perspectives were presented, too. The Confucian ethics of obligation and deferential respect offers a contrast to European–American egalitarianism and self-determination. Levinas offers a radical departure of focus from "I" to "the other" and the inherent indeterminacy of ethical interactions. Gert offers a thoughtful integration of traditional notions into a contemporary framework.

The theories presented here reflect a process of selection, of course. This selection, it is hoped, will prove informative and useful for personal reflection and social discussion. It is intended to capture most of the notions that we use in our everyday conversations about ethics. This common usage reflects their enduring relevance and usefulness, but this popularity does not diminish them. As our examination of the sophisticated philosophies behind these notions has shown, these notions address deep, perennial concerns about how best to live our lives. There are many other perspectives that we do not have space to consider here. These include various religious ethics such as those of Islam and Christianity; environmental ethics preserving the environment; political ethics of many sorts including Marxist ethics that critique the tacit values of technological, industrial cultures; situational ethics; and many others. You are urged to explore these other ethics to further cultivate your own ethical judgment and to expand your ethical awareness.

Topics for Papers and Discussion

In developing any of the following topics, you might wish to go beyond the suggested readings. You can of course locate excellent sources on your own, or you can rely on literature reviews by others. In an earlier book I wrote, *Humanistic Aspects of Technical Communication*, the chapter on Ethics surveys and categorizes most of the publications on ethics related to technical writing and communication from about 1970 through the early 1990s. In that same book, the chapter "Feminist Critiques of Science and Gender Issues" reviews the theories and several major publications in this area through the early 1990s.

1. Many contemporary thinkers point out that ethics and rhetoric play important roles in science as a social enterprise. Dale Sullivan has written extensively in this area. Much of what he writes about science applies equally well to technology and technical communication. Read his "Epideictic Rhetoric of Science," which links the rhetoric of science to its values. (In classic Aristotelian terms, rhetoric occurs in three forms: forensic, dealing with judgments about past actions such as crimes; deliberative, dealing with making laws and deciding policies; and epideictic, dealing with the celebration and propagation of important societal values.) Sullivan shows how thoroughly rhetorical and ethical scientific discourse is by revealing how the various functions of epideictic rhetoric are fulfilled in scientific discourse. These same principles and observations apply to technology and technical communication.

Summarize this article in a brief report. Think of instances from your own reading, experience, or education that illustrate the same points that Sullivan makes and include them in your report. Alternatively, you might offer illustrations that differ from Sullivan's analysis.

2. Feminist ethical critics point out that technology amounts to a value system. As a value system, it predominates in our culture, and all too often blocks out other value systems from our awareness. Beverly Sauer has written several articles on the role of exclusively objective, concrete values in the technology for investigating disasters in the mining industry. This overemphasis on technical expertise and on objective data block out subjective intuitions, leading to the silencing of important and sensible voices and the disregard of nontechnical knowledge. It can also lead to incorrect and misleading conclusions about the causes of mining disasters. Read Sauer's articles "Sense and Sensibility in Technical Documentation: How Feminist Interpretation Strategies Can Save Lives in the Nation's Mines" and "The Engineer as Rational Man: The Problem of Imminent Danger in a Non-Rational Environment." Report on Sauer's findings and conclusions. Be sure to indicate how values are reflected in the rhetoric of this technical discourse. Suggest some ways in which her conclusions might apply in other technical or scientific situations you are familiar with. You might also want to interview experienced technical communicators on this topic.

3. Values, whether implicit or explicit, are reflected in all manner of rhetoric, as we have seen. Lisa Tyler's "Ecological Disaster and Rhetorical Response: Exxon's Communications in the Wake of the Valdez Spill" shows how values get communicated in discourse, whether intentionally or unintentionally. In this case, communications seem to have been crafted without conscious consideration of the values behind the text. The result was serious damage to the credibility and reputation of Exxon in the public's eyes. Read Tyler's article and write a brief report summarizing the interplay between rhetoric and ethics as values. Analyze this discourse from several perspectives such as the Kantian or the utilitarian.

4. Ethics has to do with making choices responsibly. These choices can be large scale or small scale. Sam Dragga's " 'Is This Ethical?" A Survey of Opinion on Principles and Practices of Document Design" presents an empirical study using specific small-scale alterations to documents to pin down how teachers and professional technical communicators actually judge ethics in practice. Most people agree on what is ethical or not and focus on the consequences of these alterations in making their judgments. Read and summarize this article, being sure to identify the conclusion Dragga draws for the profession. Further establishing the need for ethical guidance in our profession is Dragga's "A Question of Ethics: Lessons from Technical Communicators on the Job," Dragga's study of 48 professional technical communicators and how they go about making ethical decisions as they work. What he found has important implications for why and how ethics might be taught in the technical communication classroom.

Write a report on Dragga's research, including his historical review, methodology, results, conclusions, and the application of these conclusions in teaching. Dealing as we do with technology every day, we might be inclined to think of ethics as a sort of technology that could be reduced to concrete rules that can be applied mechanically. What would Dragga say about such an idea?

Alternatively, you and your group could conduct your own small survey of technical communication practitioners or faculty on a specific topic of ethical concern to you and your group.

5. Sophisticated technologies are becoming more widespread, and modern business is becoming more global. As a result, technical communications is becoming more international and so needs to become more sensitive and responsive to different cultures and their values.

Read several articles on cross-cultural technical communications and report on what they say about values and ethics specifically. You might look at Edmond H. Weiss's "Technical Communication across Cultures" and Beverly Sauer's "Communicating Risk in a Cross-Cultural Context," but other sources can be readily found. Evelyn P. Boyer and Theora G. Webb's "Ethics and Diversity: A Correlation Enhanced through Corporate Communication" examines values across cultures. In your report, discuss ways in which various value systems are similar as well as different. You might want to include any cross-cultural issues you are aware of from your own experience.

6. Ethics and the law are similar in representing values. The law is more binding and codified than ethics and more stringent and more explicit in its expectations. The law is also grounded in language, whether orally in ancient cultures or in written codes in modern culture, and in the complex body of interpretations associated with the law.

In technical communications we need to be careful to warn against all the possible dangers that might follow from our documents, for both ethical and legal reasons. Howard T. Smith and Henrietta Nickels Shirk's "The Perils of Effective Documentation," though it addresses legal issues of product liability, can also serve as a guide to our ethical "liabilities" or responsibilities in technical communication. Another article that explores the legal dimensions of rhetoric in technical communication but also with clear ethical implications is Gerald M. Parsons' "A Cautionary Legal Tale: The Bose vs. Consumers Union Case." Another study of the legal consequences of the rhetoric of technical communication with clear implications for ethics is Michael J. Zerbe, Amanda J. Young, and Edwin R. Nagelhout's "The Rhetoric of Fraud in Breast Cancer Trials: Manifestations in Legal Journals and the Mass Media—and Missed Opportunities."

Read one or two of these articles or others and summarize the connections between the language, rhetoric, and content of the document and the values and responsibilities it should carry out. Include, if you can, other illustrations of the same points from your own experience or reading. How would some of the perspectives we have examined in this chapter, such as the Kantian or feminist, come into play in technical communications such as these?

7. We have studied four major ethical perspectives. Select several popular technical communication textbooks and determine how they treat the subject of ethics. Some textbooks base their discussion on theory, whereas others do not.

Some treat ethics in a very broad sense covering social concerns such as sexual discrimination and environmental protection, whereas others take more limited views concerning accuracy and clarity. Explain in what ways the four perspectives we have studied are exemplified in the textbooks you selected and in what ways they are not.

8. Confucian ethical scholarship is built upon the study of exemplary people and of commentaries on their actions. Therefore ethical texts are collections of episodes, examples, and commentaries that are not meant to be understood sequentially like a logical argument; some examples even appear to conflict with others. This basically amounts to a casebook approach to ethics, providing rules by examples. This casebook approach carries over into all aspects of Chinese business, furthermore. Study and make an oral report on a book on business negotiations with the Chinese to see the importance of ritual social courtesies, the complexity of interrelations among professionals representing corporations, and the ethical values operative in these negotiations. As suggestions, see Carolyn Blackman's *Negotiating China: Case Studies and Strategies* or Laurence J. Brahm's *Negotiating in China: 36 Strategies.*

9. As a term project in small groups, research the ethical perspective of another non-European–American culture such as Buddhism. What are the similarities and differences between it and some of the traditional Western perspectives we have examined? What would be some of the possible advantages and disadvantages of this ethical perspective for a typical technical corporate context in the United States?

REFERENCES

Allen, Jo. "Gender Issues in Technical Communication Studies." *Journal of Business and Technical Communication* 5/4 (October 1991): 371–92.
Aristotle. "Nicomachean Ethics." *Introduction to Aristotle.* Ed. Richard McKeon. New York, New York: Modern Library, 1992.
Belenky, Mary F. *Women's Ways of Knowing: The Development of Self, Voice, and Mind.* New York, New York: Basic Books, 1986.
Blackman, Carolyn. *Negotiating China: Case Studies and Strategies.* St. Leonard's, NSW, Australia: Allen & Unwin, 1997.
Bleier, Ruth. *Feminist Approaches to Science.* New York: Pergamon Press, 1986.
Bowden, Peta. *Caring: Gender-Sensitive Ethics.* London and New York: Routledge, 1997.
Boyer, Evelyn P. and Theora G. Webb. "Ethics and Diversity: A Correlation Enhanced Through Corporate Communication." *IEEE Transactions on Professional Communication* 35/2 (March 1992): 38–45.
Brahm, Laurence J. *Negotiating in China: 36 Strategies.* Singapore: Reed Academic Publishing Asia, 1995.
Brasseur, Lee. "Contesting the Objectivist Paradigm: Gender Issues in the Technical and Professional Communication Classroom." *IEEE Transactions on Professional Communication* 36/3 (September 1993): 114–23.
Burke, Kenneth. *Language as Symbolic Action.* Berkeley, California: University of California Press, 1966.

Daly, Mary. *Gyn/Ecology: The Metaethics of Radical Feminism.* Boston, Massachusetts: Beacon Press, 1990.

Dawson, Raymond. *Confucius.* New York: Hill and Wang, 1981.

Dombrowski, Paul M. "Ethics." *Humanistic Aspects of Technical Communication.* Amityville, New York: Baywood, 1994.

———. "Feminist Critiques of Science and Gender Issues." *Humanistic Aspects of Technical Communication.* Amityville, New York: Baywood, 1994.

———. "Can Ethics Be Technologized? Lessons from *Challenger,* Philosophy, and Rhetoric." *IEEE Transactions on Professional Communication* 38/3 (September 1995): 11–16.

Dragga, Sam. "Women and the Profession of Technical Writing: Social and Economic Influences and Implications." *Journal of Business and Technical Communication* 7/3 (July 1993): 312–21.

———. " 'Is This Ethical?' A Survey of Opinion on Principles and Practices of Document Design." *Technical Communication* 43/1 (First Qtr. 1996): 29–38.

Etz, Donald V., "Confucius for the Technical Communicator: Selections from *The Analects.*" *Technical Communication* 39/4 (1992): 641–644.

Gert, Bernard. *Morality: A New Justification of the Moral Rules.* Oxford, England: Oxford University Press, 1988.

Gilligan, Carol. *In A Different Voice: Psychological Theory and Women's Development.* Cambridge, Massachusetts: Harvard University Press, 1982.

Guthrie, W. K. C. *The Sophists.* Cambridge, England: Cambridge University Press, 1971.

Hansen, Chad. "Classical Chinese Ethics." *A Companion to Ethics.* Ed. Peter Singer. Oxford, England: Blackwell, 1991.

Harding, Sandra. *The Science Question in Feminism.* Ithaca, New York: Cornell University Press, 1986.

———. Harding, Sandra. *Whose Science? Whose Knowledge?* Ithaca, New York: Cornell University Press, 1991.

Held, Virginia, Ed. *Justice and Care: Essential Readings in Feminist Ethics.* Boulder, Colorado: Westview Press, 1995.

Hirsch, Marianne, and Evelyn Fox Keller, Eds. *Conflict in Feminism.* New York, New York: Routledge, 1990.

Kant, Immanuel. *The Fundamental Principles of the Metaphysic of Ethics.* Otto Manthey–Zorn, Trans. New York, New York: Appleton–Century–Crofts, 1938.

Katz, Steven B. "The Ethic of Expediency: Classical Theoretic, Technology, and the Holocaust." *College English* 54/3 (1992): 255–75.

Keller, Evelyn Fox. *A Feeling for the Organism: The Life and Work of Barbara McClintock.* San Francisco, California: W. H. Freeman, 1983.

Kimball, Bruce A. *Orators and Philosophers.* New York, New York: Teachers College, Columbia University Press, 1986.

Kolodny, Annette. *The Lay of the Land: Metaphor as Experience and History in American Life and Letters.* Chapel Hill, North Carolina: University of North Carolina Press, 1975.

Larrabee, Mary Jeanne, ed. *An Ethic of Care: Feminist and Interdisciplinary Perspectives.* New York, New York: Routledge, 1993.

Lay, Mary. "Interpersonal Conflict in Collaborative Writing: What We Can Learn from Gender Studies." *Journal of Business and Technical Communication* 3/2 (1989): 5–27.

———. "Feminist Theory and the Redefinition of Technical Communication." *Journal of Business and Technical Communication* 5/4 (1991): 348–70.

Lessl, Thomas. "The Priestly Voice." *Quarterly Journal of Speech* 75 (1989): 183–97.

Levinas, Emmanuel. *Collected Philosophical Papers.* Trans. A. Lingis. Dordrecht, The Netherlands: Martinus Nijhoff, 1990.

Lloyd, G. E. R. *Aristotle: The Growth and Structure of His Thought.* London, England: Cambridge University Press, 1968.

Markel, Mike. "A Basic Unit on Ethics for Technical Communicators." *Journal of Technical Writing and Communication* 24/4 (1991): 327–50.

Mill, John Stuart. *Utilitarianism*. Ed. Oskar Piest. Indianapolis, Indiana: Bobbs–Merrill, 1957.

Miller, Carolyn R. "Technology as a Form of Consciousness: A Study of Contemporary Ethos." *Central States Speech Journal* 29 (Winter 1978): 228–236.

Noddings, Nel. *Caring: A Feminine Approach to Ethics and Moral Education*. Berkeley, California: University of California Press, 1984.

Parsons, Gerald. "A Cautionary Tale: The Bose vs. Consumers Union Case." *Journal of Technical Writing and Communication* 22/4 (1992): 377–86.

Rorty, Richard. *Objectivity, Relativism, and Truth, Vol. I*. Cambridge, England: Cambridge University Press, 1994.

Sauer, Beverly. "The Engineer as Rational Man: The Problem of Imminent Danger in a Non-Rational Environment." *IEEE Transactions on Professional Communication* 35/4 (1992): 242–49.

_____. "Sense and Sensibility in Technical Documentation: How Feminist Interpretation Strategies Can Save Lives in the Nation's Mines." *Journal of Business and Technical Communication* 7 (1993): 63–83.

_____. "Communicating Risk in a Cross-Cultural Context." *Journal of Business and Technical Communication* 10/3 (July 1996): 306–30.

Smith, Howard T., and Henrietta Nickels Shirk. "The Perils of Defective Documentation." *Journal of Business and Technical Communication* 10/2 (April 1996): 187–203.

Sturges, D. L. "Overcoming the Ethical Dilemma: Communication Decisions in the Ethic Ecosystem. *IEEE Transactions on Professional Communication* 31/1 (1992): 44–50.

Sullivan, Dale L. "Political-Ethical Implications of Defining Technical Communication as a Practice." *JAC: Journal of Advanced Composition* 10 (1990): 375–86.

_____. "The Epideictic Rhetoric of Science." *Journal of Business and Technical Communication* 5/3 (July 1991): 229–45.

Tong, Rosemarie. *Feminine and Feminist Ethics*. Belmont, California: Wadsworth, 1993.

Tuana, Nancy. *Feminism and Science*. Bloomington, Indiana: Indiana University Press, 1989.

Tyler, Lisa. "Ecological Disaster and Rhetorical Response: Exxon's Communications in the Wake of the Valdez Spill." *Journal of Business and Technical Communication* 6/2 (April 1992): 149–71.

Wei-ming, Tu. "The Confucian Tradition in Chinese History." *Heritage of China: Contemporary Perspectives on Chinese Civilization*. Ed. Paul S. Ropp. Berkeley, California: University of California Press, 1990.

Weiss, Edmond H. "Technical Communication Across Cultures." *Journal of Business and Technical Communication* 12/2 (April 1998): 253–70.

Wicclair, Mark R. and David K. Farkas. "Ethical Reasoning in Technical Communication: A Practical Framework." *Technical Communication* 31/2 (1984): 15–19.

Zerbe, Michael J., Amanda J. Young, and Edwin R. Nagelhout. "The Rhetoric of Fraud in Breast Cancer Trials: Manifestations in Medical Journals and the Mass Media—and Missed Opportunities." *Journal of Technical Writing and Communication* 28/1 (1998): 39–61.

CHAPTER 4

NAZI RECORDS
THE ORIGIN AND USE
OF INFORMATION

This chapter deals with ethical issues about how technical information is obtained and how it will be used. Understanding these issues will help us to consider whether or how we should be communicating the information we deal with. We will be asking the following questions in this chapter: Should it make any difference to us as communicators how a particular body of information that we might communicate was obtained? How are ethics and values reflected in our language and in the format of our technical documents? Should the uses to which our information will likely be put influence our ethical judgment about the communication of that information? Grappling with these questions will help us to see that information, whether scientific or technical, often does not exist as an isolate in crisp purity all by itself but is usually embedded in complex contexts. These contexts establish where the information comes from, how and why it came to be discovered, and what will become of it as it is used. These contexts, whether social, cultural, political, scientific, or technological, are essentially human, involving real lives in all sorts of ways.

We will be examining several very dramatic and appalling instances of technical information generated by the notorious Nazi regime of Germany in the late 1930s to 1945. The ethical issues about the communication of technical information from the Nazis are presented here because they represent as starkly as possible their particular points, points that in less dramatic circumstances might seem obscure or unimportant. The purpose of citing these communications is only to illustrate some very real potential ethical pitfalls—pitfalls that we, in lesser degrees, might confront in real technical communication situations in our times. We will see, for example, how similar issues surround recent revelations about U.S. research on the effects of radiation on humans in the decades just after World War II. We will also see how the U.S. Environmental Protection Agency dealt with similar issues just recently.

Technical communication is always a very human activity, people connecting with other people about matters of mutual human concern. It involves perhaps the most important distinguishing features of humankind: language and the sense of ethical responsibility. The examples we will examine show one way ethical considerations can be abandoned in communications—with momentous consequences. The lesson of these examples is that an excessive emphasis on the values of technical objectivity (as well as scientific), technical excellence, or technical expediency can sometimes mask vital ethical issues.

As you read this chapter, keep in mind that these examples are not attempting somehow to explain the Nazi regime and the way it treated people as objects in their technical communications. As many scholars have pointed out, when one tries to explain something, one is often also trying to justify it. For these scholars, though, the Nazi regime is beyond explanation and justification, utterly beyond the understandable. Nevertheless, we can learn important ethical lessons even from what is not fully understandable; indeed, part of our ethical responsibility is to learn from the past and, in this case more than any other, never to forget.

We will examine issues about information obtained from "scientific" research on prisoners. The term "scientific" is in quotation marks here for two reasons. First, it indicates the explicit claim by the researchers that the research was scientific in nature. Second, and more important, it indicates that nearly all of this research is now widely understood by the legitimate scientific community as being completely without genuine scientific purpose or validity. Its aim was, instead, principally racial abuse and mass killing, not the pursuit of science, regardless of any claims to the contrary.

The point of these examples is to show that some of the values embedded in the scientific frame of mind can be carried to extremes—with terrible consequences. These same values for the most part are operative in the advancement of technology, too. Of particular concern are the emotional disengagement of the researcher from the human research subject, the great power and control differential between them, and the implicit superiority of the researcher by which the subject is deliberately kept in the dark about what is really going on. These values have always held an important place in scientific inquiry. They can, however, be carried to extremes. We should always therefore be alert to carrying technical and scientific objectivity too far.

As technical communicators, we are used to thinking of information as factual knowledge that exists apart from values, politics, or racial biases. We are also used to thinking of subject matter experts as responsible for any ethical dilemmas involving the technical information they generate. Our role, we might think, is only to communicate that information effectively. This leaves the ethical burden with others, either the originators or the end users. The examples we will examine, however, show that technical communicators and the technical documents we produce are not as ethically neutral as we might think.

Science, or technology, as a value system—its own ethical universe so to speak—is woefully impoverished, as the critics we encounter throughout this book repeatedly point out. Steven Katz, Carolyn Miller, Dale Sullivan, Beverly Sauer, Mary Lay, and other technical communication scholars say that all too easily the narrow, instrumental values of technical expediency or technical excellence can rise to the forefront and be taken as the only relevant values in a given situation. To do so excludes, however, all other values, including any feeling for the research subject as a person like yourself. Science thus becomes its own justification and warrant—science done solely for the sake of science; the same holds for technology and the pursuit of technical information.

It can be argued, however, that in the case of the Nazi "medical" research (and the same for the Nazi human extermination technology discussed later), what was going on was really not science for the sake of science. Instead, it actually was science subordinated to another value system, in this case supposed racial supremacy. Thus the medical "research" can be seen not as primarily about advancing scientific knowledge at all but rather about racial persecution. Science was just a means to accomplish the end of racial persecution and extermination. Thus it was not scientific and technological values at work but racism. This is the position of many critics.

We can see just from the preceding discussion that ethical appraisals can be quite complex and ambiguous. On the one hand, this "research" can be seen as elevating science above other values, or, on the other, it can be seen as science subordinated to other values. In either case, however, we should keep in mind that the stated justification for acquiring this information was to advance medical knowledge. As we will see (in the Lifton book discussed later), the highest level of the Nazi regime mandated that doctors be involved intimately in the execution process in order to lend the appearance of scientific and medical legitimacy and authority to these terrible deeds.

It is easy to criticize the Nazi regime and its work. It was radically unethical and constitutionally illegal. It was also so profoundly evil that legal scholars had to create a whole new category of crime to describe its terrible activities: genocide, the extermination of an entire race, as a crime against humanity itself. It is therefore perhaps too easy an ethical target. Our condemnation can feel so certain and so sweeping that we can end up losing our focus on technical communication specifically. We can readily abhor an entire regime and an entire war. But life is lived individually, and it is each separate event and individual life that must be remembered and appraised. Therefore my immediate purpose here is to focus on only a few specific examples to illustrate several specific ethical points, not on a global condemnation of the regime, which others have done so well already.

In the first example, we examine the recent controversy about whether information obtained illegally and unethically by Nazi "researchers" should be disseminated and used at all. This controversy shows the need to examine how information originated in order to fully understand all its ethical dimensions. It shows too how problematic real ethical questions can be, with two opposing sides making reasonable arguments to support their positions.

In the second example, we examine a specific Nazi technical document. This document, posted on many Holocaust Web sites and already well analyzed in a publication by Steven Katz, clearly illustrates the interplay between values and language. Not only does this document exemplify certain values in itself but it also puts into concrete action a whole set of values from its cultural context. Many of these values deal with the uses to which the information in the document will be put. It also illustrates well some of the key values of both technology and science as social enterprises: objectivity, emotional distance, indifference to persons, and instrumental effectiveness. As a whole, our ethical analysis of this document clearly demonstrates the need to situate technical communication within its social and cultural context in order to understand its full ethical significance.

ORIGINATION, DISSEMINATION, AND USE OF INFORMATION

In this section we briefly review the Nazi regime in its racism, its relation to science and medicine, and its trial for crimes against humanity. We then review recent ethics debates about questionable scientific information from this regime. The purpose of these reviews is to show that ethical considerations apply not only to the document itself or its content but also to how the informational content was obtained and how it likely will be used. Thus the lessons from this past regime apply to contemporary technical communication, too.

Bernadette Longo has pointed out that technical communications are always implicated in many ways in the prevailing culture, whether it is the culture of the individual business you happen to be writing for, or the entire industry, or the larger society. All these groups have interests being played out in any given technical document. Because they define where the document came from and why it was produced, it is impossible to fully understand the document, especially its ethical dimensions, without a full understanding of this multifaceted context as well.

First we will review the Nazi regime as a historical fact. Next we will examine several recent controversies about scientific and pseudoscientific information from the Nazi regime. After that we will examine the traditional view about the values behind Nazi medical sciences as an ethic of sorts driving its research. We will see that objectivization (the excessive and inappropriate treatment of people as objects), impersonalness, and emotional disengagement are key values in Nazi medical science. These values, in addition to others such as instrumental effectiveness, are cited by modern critics as being highly problematic and cause for serious concern. These values are commonly attributed to modern technology, too, so these concerns apply as well for contemporary technical communication. We will then consider an alternative view of the values behind Nazi medical science, one that shows it to be really antiscience rather than science.

In the last section, on suggested discussion and paper topics, some of these concerns are seen to parallel issues about technical and scientific matters in contemporary America. These matters include research on the human effects of radiation and concerns by the government for ethical standards in biomedical research.

Nazi Past

Before examining specific instances, we need to understand the historical context. Following the end of World War II in Europe, many Nazi leaders were put on trial for war crimes to civilians as well as to soldiers. These were the famous Nuremberg trials, notable for revealing the need for a totally new term, genocide, for the immense Nazi crimes against humanity and the Jewish people. Among these war criminals were leaders of the infamous concentration camps and death camps.

The people held in these camps included a few common criminals, but by far the greater number were those who were simply thought to be undesirable by the Nazi regime but who were not guilty of any crime at all. These included Jews of various nationalities, who were persecuted for their race and religion. They also included non-Jews of Slavic and other ethnicities held to be inferior to the "Aryan" race from which native Germans supposedly originated. Still others included homosexuals, gypsies, and the weak and infirm, regardless of their nationality or ethnicity. Many of these camps were concentration camps meant to isolate undesirables from the general populace. From there some people were sent to forced labor under miserable conditions in various industries as what amounted to slaves.[1] Some were retained as subjects for various supposedly scientific research experiments, which often ended in death. Still others were simply held until they could be killed.

The Nuremberg trials are notable for many shocking revelations, one of which was the horrendous institutionalized abuse of people in the name of scientific and medical research. Most of this so-called scientific research was intended to support the Nazi war effort by showing the limits of human endurance and so amounts to a sort of "technical" research, too. In particular, the trials show that the emotional distance and control that the researcher has over the research subject, who could be treated as an object, holds great potential for abuse. As a result of these revelations (and other factors), scientific study in Europe and America has been closely scrutinized to prevent anything remotely like the Nazi pseudo-research from ever happening again. For that reason, in our universities today all scientific research involving people as research subjects has to be reviewed by "human subjects" panels. These panels must include nonscientists who are in no way connected with the research or its potential uses. This is true not only for research in physiology and medicine but even in psychology and education.

---

[1]See for example *The Rocket and the Reich*, which shows the brutal use of slave labor in manufacturing the V-2 rocket weapon. In this way the setting of "pure" science and technology in which Werner von Braun and other rocket scientists are usually portrayed is shown to be anything but pure. It was heavily tainted with human abuse, death, militarism, and racism. Thus high technology was made possible by slave labor.

Controversy in the Present

In recent years a new outcry has arisen in several different forms about Nazi "scientific" information and other information collected unethically. This controversy has to do with medical specimens of human organs and with the dissemination and use of information obtained from unethical "research," such as the effects of phosgene, a chemical warfare agent, on humans. After considering these instances, we will examine the nature of the values behind Nazi medical "research" both from the traditional view and from a new, alternative view.

Medical Specimens.   In the first form of recent ethical concerns that we will consider, scandals have arisen over human anatomical samples used in medical education. The respected scientific journal *Nature,* for instance, reported in 1989 that Israelis were shocked and outraged to discover that tissue specimens and skeletons used for education at several prominent German medical schools, including Tübingen University and Heidelberg University, derived from Nazi prison and death camps. Though the precise circumstances of origin of the particular samples are not known, many of the deceased in question were likely Jews and had not been granted any say in whether their bodies could be used for such purposes. More important, these Jews were put to death simply because they were Jews. Sometimes prisoners were apparently even executed on demand to acquire a particular sort of tissue sample. A sample from a prison camp of a healthy organ from a young person who was not suffering from any disease, for example, is thought probably to have derived from an execution arranged to provide the desired organ.

The Israeli protests to the German government, we should note, had nothing to do with the informative value of these samples but centered on the circumstances under which they were obtained. No informed consent, no possibility of choosing otherwise, no legitimate reason for the execution, no possibility of protest—none of the familiar criteria that we consider basic human rights applied in these cases. These were instances of murder and the abuse of corpses. The Israelis insisted that these improperly obtained samples be removed from the schools and disposed of with the same respect that would be accorded an ordinary citizen. For these critics the *means* by which the samples were obtained taints them completely and should prevent them from being used for any purposes, regardless of any informative value they might hold.

"Research" Information.   A somewhat similar situation arose recently in the United States. In one case, interest arose about publishing information from Nazi hypothermia experiments in which prisoners were exposed to cold to the point of collapse, even death. This information, it was argued, could be put to use in our own times to improve survival equipment, for example. Others, however, opposed the use of this information because of the terrible, unjust suffering and persecution through which it was obtained.

The New England Journal of Medicine (NEMJ), one of the most respected American medical journals, addressed the controversy in a particularly strong way. The May 17, 1990 special issue on the controversy includes a thorough, careful appraisal by R. L. Berger of the scientific validity of the Nazi hypothermia "experiments" at Dachau as well as several other "experiments." The experiments were supposed to determine the limits of survival of pilots crash-landing in the ocean. Berger points out that in this case, as in most other ethical debates, both sides have compelling arguments. One side argues that information obtained from the camps should never be used for any purpose. The other side argues that the information should be used precisely in order to give some purpose to the victims' suffering as well as to relieve the suffering of those who might benefit from the information. Berger explains, however, that the ethical dilemma in this particular case is moot because the hypothermia research is totally unscientific—inadequate methods, unwarranted conclusions, fabricated data, and internal inconsistencies abound in the "research" materials. In another sense, though, the ethical debate was not moot. One good result of the debate has been to force an open discussion in the medical journals about the relation of knowledge to the means by which it was obtained and the ends for which it might be used.[2]

The editorial board of *NEJM* itself goes even further, exploring the ethical value of such inhuman research even if it were scientifically valid. The editors, represented by Marcia Angell, take a firm, clear ethical stand:

> The *Journal* has taken the position that it will not publish reports of unethical research, regardless of their scientific merit. Only if the work was properly conducted, with full attention to the rights of human subjects, are we willing to consider it further. The approval of the institutional review board (when there is one) and the informed consent of the research subjects are necessary but not sufficient conditions [for ethicality]. Even consenting subjects must not be exposed to appreciable risks without the possibility of commensurate benefits (1463).

Among the reasons for this position are the need to deter unethical work and the need to affirm the primacy of the research subject. (This is very much like Kant's ethics, treating others by the same rules as we would wish to be treated.) In addition, this policy "serves notice to society at large that even scientists do not consider science the primary measure of a civilization. Knowledge, although important, may be less important to a decent society than the way it is obtained" (1463–64).

Also in response to this topic, the prestigious *Journal of the American Medical Association (JAMA)* published several articles in an issue largely devoted to

---

[2]Berger notes that the complete lack of scientific validity of the hypothermia "research" was known even to Nazi leaders, who eventually had the investigator, Dr. Rascher, executed for fabricating his data.

it.[3] This special issue went beyond the particulars of specific "research" and "researchers" to examine ethically the entire social–cultural–political context that engendered these studies. Jeremiah Barondess critiques the collusion of medical doctors and the entire German medical institution of the time in this "research." He explains that a radical reversal occurred within the biomedical profession, leading it to support Hitler and Nazism, even becoming "an arm of state policy" (1657). This reversal, from medicine as healing to medicine as killing, was the result of powerful economic and political circumstances. It involved the abandonment of the Hippocratic oath with its pledge against doing harm and the abandonment of the tradition of supporting the health of the patient above all else. It involved a eugenics program geared toward racial purity (called "racial hygiene") that served to legitimate on scientific and medical grounds the state policy of racial supremacy and purification by extermination. Though this was directed primarily toward Jews, it also included the feeble and retarded as well as Slavs, criminals, homosexuals, gypsies, and dissidents, all of whom were considered racial contaminants.[4]

Barondess is clear in assigning responsibility not just to individual "researchers" but to the whole medical establishment: "Physicians lent themselves to this process in large numbers and to a degree that made much of what followed possible" (1657). Medical scientists gave scientific legitimacy to the notion that the "Aryan" race is innately superior to all others. This racism permitted treating some others as not fully human on par with "Aryans," even treating them as mere objects. It also legitimated the eugenics program, which started out with forced sterilization but eventually culminated in murder on a stupendous scale. Medical schools disseminated racial theory while doctors, as administrators and physicians, implemented the theory. Physicians, for example, presided over the executions in the death camps, which included "scientifically" selecting those to be executed from those to be preserved for slave labor, prescribing the means of their execution, and verifying the fact of their death. It was *only* because of this "scientific" sanctification by the biomedical establishment, Robert Jay Lifton explains, that these terrible activities were socially tolerated at all. Making them seem medically necessary and scientifically justified seemed to make them acceptable. Barondess draws an important ethical lesson for our contemporary world from these atrocities of the past:

> Echoes of the Nazi program continue today. They exist not only in regard to eugenics and to broader issues at the interface of medicine and the state, but more deeply in relation to the dangers of allowing clinical and research programs that reflect a hierarchy of human worth. . . . [T]he medical ethos is not immutable, but can be severely distorted by social and political forces and by perversions in the application of science and technology (1661).

[3] *JAMA*, Vol. 276, No. 20, November 27, 1996.

[4] Not all physicians participated in this role reversal, however, as Kater, Lifton, and others point out. Kohler's review of Kater offers a concise discussion on this point.

In another recent case in the United States, the U.S. Environmental Protection Agency (EPA) was considering publishing and using Nazi information obtained by subjecting prisoners to phosgene, a poisonous gas used in chemical warfare. Phosgene can also be produced in some industrial processes. The EPA was developing new standards for pollution regulation of this gas. The agency, however, had no information comparable to the Nazi "research" and no likelihood of obtaining similar information either deliberately or by accident because the Nazi data derived from closely measured doses at very high concentrations.

This "touched off a dispute among EPA scientists and others about the ethics of using the results of the Nazi experiments and, also, about the scientific quality of the studies," reported Marjorie Sun in the journal *Science* in 1988 (21). One side argued for the intrinsic usefulness of the information regardless of how it was obtained. They argued that if identical data were obtained through unintentional exposure such as in an accidental explosion at a chemical plant, the information would be considered "a gold mine" and would unquestionably have been used by the EPA. After receiving extensive protests from twenty-two EPA scientists, the EPA chief administrator, Lee Thomas, ruled to bar the use of any data obtained from Nazi experiments, regardless of the validity of their scientific methodology (which nevertheless was found to be seriously flawed).

Values in Nazi Medical "Science"

The close, mutually supportive relationship between the German medical establishment and the Nazi political regime has been explored in detail in a number of recent books. After examining the traditional view of the values behind Nazi medical research on prisoners, we will examine a new, alternative view indicating that this research was not really science but antiscience.

Traditional View. Many of these historical treatments take a sociological perspective. Michael Kater, a noted historian on the Nazi era, points out the sociological or demographic reasons that Jewish doctors were scapegoated for the general troubles of German society of the time. He also notes that physicians were the most Nazified of the professions, being overrepresented in the party by a factor of seven compared to most other professions. The glut of medical school graduates and their underemployment in society, the depressed economy, and the need for the Nazi regime to legitimate its racial policies were all important factors at work to medicalize the mistreatment of prisoners and patients, Kater says.

Another book, by Robert Jay Lifton, also a well-publicized historian on the Nazi era, has drawn considerable attention. It is directly relevant to technical communication because it clearly outlines some of the uses of language and communication in support of the Nazi regime's plans for its victims. Lifton's primary focus is the paradoxical inversion of the very meaning of what it is to be a doctor and of the purpose of the medical establishment. Put simply, the healer became the killer, and healing became killing.

Such a stark and absurd reversal of meaning, driven by contextual circumstances, has appeared in other situations relevant to technical communication. As the chapter on the shuttle *Challenger* disaster shows, indications of danger were somehow transmuted into indications of safety; in effect, "black became white." In that case as well as this, one lesson we should draw from such events is to understand the powerful social contingency of meaning. Rather than accept technical facts and factual statements as absolutely true, unalterable and incontrovertible, we should be aware of how readily they can be transformed by social circumstances. For this reason ethics in technical communication must always be tied to rhetoric as persuasion and the negotiation of social consensus and to social constructionism as the social ratification of factual statements.

Lifton points out that, perversely, the execution of numberless Jews and non-Jews was legitimated as medically necessary for the sake of racial health and purity. The killing was further justified, in light of the war effort, on the basis of the need to conserve precious resources of food, medicine, hospital beds, and staff, which were needed for more worthy individuals, namely soldiers. But the war effort was aimed at the same goal as the medical killings—the eradication of all Jews everywhere. Thus, the war supported the killing and not the other way around, as the regime claimed.

Doctors controlled the killing from start to finish from the earliest days until the end of the war. Communications directing the medicalization of killing came from the top, Hitler, and were reinforced progressively throughout the medical bureaucracy. In the early days of the killing, only the most helpless and vulnerable were put to death, namely institutionalized mentally and physically disabled children. Doctors identified them, certified their disabilities, and supervised their execution by various means, such as starvation.

Lifton emphasizes that the legitimation of these procedures hinged on the approval of the medical establishment at the highest level down to the lowest. Most of these doctors were also members of the dreaded SS. In hospitals, children were executed under doctors' orders and often in their presence precisely in order to legitimate these activities. Afterward, doctors confirmed that death had occurred and wrote false death certificates to relatives. In the death camps later, Lifton explains, upon disembarking from the trains at camp, prisoners were separated into those to be gassed immediately and those to be retained for slave labor or medical experiments. Lifton says:

> In terms of the actual professional requirements, there was absolutely no need for doctors to be the ones conducting selections: anyone could have sorted out the weak and moribund prisoners. But if one views Auschwitz, as Nazi idealogues did, as a public health venture, doctors alone became eligible to select. In doing so, the doctors plunged into what can be called the *healing-killing paradox* (150).

One of the lessons here for technical communication is that an inversion, or at least a radical reinterpretation, of meaning requires a complex chain of circumstances and participants to affirm that meaning. The interrelatedness of the links of the chain, furthermore, allows any one individual readily to point to the entire rest of the chain, up and down, to justify one's own complicity. For example, many doctors, Lifton learned through interviews with them, understood the patients they selected to be executed as "already dead," as determined by the medical system, and so their own involvement in the execution was a minor step in an implacable chain of events. The very terminology was involved in this process of the reversal or inversion of meaning. The chilling notion of the "already dead" was coupled with the medicalized racist concept of the disabled and infirm as "life not worthy of life."

Masked language also played an important role in communications about the medical killings in many ways, both externally to the public and internally to bureaucrats, military officers, and doctors. This masked language allowed one to avoid expressing one's full meaning clearly and unequivocally so as to avoid arousing public outcries and to avoid taking ethical responsibility. It also allowed a broad range of interpretations that could conveniently serve one's own interests to try to demonstrate that one did not know what was really going on. The most obvious of these instances of masked language were "euthanasia" and "special treatment."

"Euthanasia" historically has been understood as mercy killing, the putting to death of someone of sound mind with a terminal condition who is in such unrelenting extreme misery that he or she wants life to end sooner rather than later. The knowing, express wish of the person has always been a prerequisite of euthanasia. In the Nazi regime, however, euthanasia was reinterpreted as putting someone to death in a way that was perceived as humane and on the basis of his or her unworthiness to live, according to the perceptions of the regime. Thus children could be gassed with carbon monoxide or fed a large dose of a powerful sedative to bring about death. This was a "happy death," euthanasia, supposedly because it was not violent, in contrast to, say, a bullet to the back of the neck. As the medical executioners quickly learned, few had the stomach or heart to kill children with bloody violence, but many could without qualms put children to sleep.

It was no coincidence that the killing began with institutionalized disabled children, Lifton explains. These people could not protest and were isolated from society and were therefore unseen and unnoticed. As time went on, the scope of the killing widened to include disabled adults, both Jew and non-Jew, and those perceived as problems to or burdens on society—criminals, prostitutes, homosexuals, the chronically unemployable, and many others. As numerous critics have pointed out, historically the primary purpose of the law in any society has always been precisely to protect those who cannot protect themselves. In this case, however, the law, authorized by the medical establishment, served precisely the opposite purpose.

The inversion of healing to killing was justified on the basis of a further paradoxical reinterpretation of meaning. These killings were understood as justified for the health of the Germanic (i.e., "Aryan") people, who were considered to be superior to all other people and nations. For the sake of the purity and health of this organism, the Germanic people, all other people were considered as potential contaminants of the purity and therefore health of the Germanic people. The Jews in particular were singled out as the most dreadful contamination, which had to be utterly eradicated. The medical analogy was repeatedly used to justify this policy, Jews being compared to gangrenous limbs or to cancers or to vermin as transmitters of disease. As any public health official knows, disease and the sources of disease must be eradicated. For this reason, doctors were totally involved as public health officials in a commanding role in the killing.

Another inversion and evasion of meaning was "special treatment." Ordinary medical problems are amenable to ordinary medical treatments. The most severe, virulent, and dangerous medical problems, however, by their special nature require special treatments. Typically, "special treatment" would be understood to mean unusually earnest and powerful treatment, still with the aim of preserving the patient. In this case, however, "special treatment" referred to medical killing, special in the sense of lying outside the mainstream of medicine as traditionally understood. Lifton observes: "The word [*special*] not only detoxified killing and aided in its routinization but, at the same time, infused that killing with a near-mystical priority. . . . Killing assumed a certain feeling of necessity and appropriateness, enhanced by the medical, as well as the military, aura surrounding it" (150–151).

Regarding the alleged scientificness of the horrifying experiments performed on prisoners, there is practically none, Lifton points out. From interviews with prisoner–doctors who were forced to participate in these "experiments," he concludes that they were done either to satisfy psychopathic cruelty or personal aggrandizement, or to attempt to support Nazi racial theories, or to lend support to the medical-killing paradox that the killings were medically and scientifically necessary.

In addition, besides justifying themselves to the regime and the regime to the public, these medical "experiments," Lifton concludes, served as a psychological defense and rationalization for the doctor researchers. "He [the doctor researcher] and other doctors could follow a pattern . . . : the immersion of themselves in 'medical science' as a means of avoiding awareness of, and guilt over, their participation in a murderous project" (61).

Nazi Antiscience.   Many explanations have been offered for the horrible activities that constitute this complex puzzle. Of course, a vicious racism underlies whatever other factors are offered as explanations. Conformity to the wishes of political leaders or to governmental policy plays a role, too, especially in the form of institutionalized racism. In addition to these sociological views, a personal, psychological view further complicates the puzzle. Sadism and plain cruelty are often also invoked as explanations. Revenge for perceived injuries and psychological compensation for perceived inadequacies played important roles, too, others argue, especially for those pseudoresearchers who were otherwise unable to achieve publication in respected academic journals.

For some critics, these horrible activities done in the name of medical science are explained on the basis of the intrinsic inhumaneness and unethicalness of science itself. These critics perceive these Nazi activities as extreme forms of the basic principles underlying all science and technology—a specific instance of a general problem. For the rhetorician and ethicist Richard Weaver, for instance, these activities reflect the impersonalness that supposedly is a cornerstone of the scientific enterprise. These actions thus show science to be essentially unethical and an enemy of basic human values. Other ethical critics of science and technology, such as Jacques Ellul and Langdon Winner, are likewise concerned about the impersonalness of science and technology. They are also greatly concerned about the apparent indifference of science and technology to traditional values and social goals. They wonder about the desirability of allowing science and technology to pursue their unbridled courses in our society wherever they might lead. Stephen Monsma has raised similar concerns about our responsibility for the ethical stewardship of the earth, to protect it from long-term pollution. Leo Marx, similarly, wonders about the unbridled growth of science in pursuing whatever course it may.

Carolyn Miller, the rhetorician and humanist, is also concerned about the "objectivization" that goes hand-in-hand with progress in modern science and technology. Objectivization means treating people as objects rather than as persons equivalent in every way to the researcher. Because it is fundamentally dehumanizing, objectivization is potentially dangerous and should be restrained by traditional humanistic interests, which insist on open criticism and ethical appraisal. More strongly, many feminist critics of science such as Beverly Sauer and Lee Brasseur pointedly oppose the dehumanizing that seems to go along with modern science and technology. This dehumanizing is typically done on the disadvantaged and less powerful people in society—the sick, weak, poor, and powerless.

We should keep in mind that to explain this Nazi pseudoresearch in terms of science or technology requires some serious assumptions about the nature of science and technology. It assumes, principally, that science and technology are at best indifferent to or even oppose conventional human values. And we know that, from an ethical point of view, indifference amounts practically to the abandonment of ethical principles. Outright opposition to conventional values, of course, is even more dangerous. In the case of the Nazi doctors, it appears that much of this research was done in the death camps *because* it could not have been conducted in conventional medical settings under conventional ethical guidelines. It was as though the opportunity had to be seized precisely because it circumvented traditional values.

The true situation is even more complex, however. Keep in mind, as we saw earlier, that ethics is innately problematic and rarely yields simple, easy answers. Perhaps the concerns just mentioned about science are not entirely on the mark for the Nazi regime. Anne Harrington, professor of the history of science at Harvard, notes that most of the critical commentary about the Nazi regime's medical pseudoresearch has characterized it as "objectivity run amok."

This characterization explains that "the most pernicious energy driving the engine of Nazi medical science had *not* been racism, anti-Semitism, political agenda, and so on; it had been its perverse fidelity to an obscene *objectivity* that ultimately found it possible to see all activities through the lens of expediency, scientific 'interest,' and efficiency" (184). Of course, this view assumes that objectivity, and science with it, is innately dehumanizing, "an epistemological stance that denied scientific objectivity the capacity to *see* humanely" (184). Harrington cites another critic as saying about the unethicalness of science, " 'Scientists espouse objectivity and spurn value judgments. But pure objectivity leads to regarding everything as being feasible. . . . For these scientists, objectivity opened the door to every conceivable form of barbarism' " (in Harrington, 184–85).

Harrington herself, however, explains that this perspective is flawed largely because it reflects the commentators' preconceptions, which operate backward in time to yield post facto judgments of questionable validity:

> Is it an accident that such stories get written and read with such passion in today's era of high-tech medical treatment, relentless research, and fears that the voice of patients and the ethic of personal clinical care are being lost in a system that values "cure" as a dissociated end in itself? In other words, is it possible that the story of Nazi biomedicine as "objectivity run amok" resonates so readily for critics of modern biomedicine, not only because it is true on its own terms, but because the writers involved in the present-day critique already "know," from their own experiences, how true this particular reading of the past "must" be? (185).[5]

Harrington's own investigation, together with that of several others, reveals that actually the Nazi regime and these pseudoresearchers were operating under exactly the opposite frame of mind. Those researchers were pursuing not science but what amounts to antiscience. Harrington explains that many medical researchers, and a good deal of the general populace in Nazi Germany, were disenchanted with traditional empirical science, so disenchanted as to deliberately, specifically oppose traditional science. These people were disenchanted by its rigorous objectivity, its formal logic, its emotional (and political) neutrality, its mechanistic and reductive view of people, and its analytical method that breaks things into their parts rather than preserving wholes. These antiscientific investigators, rather than insisting on the objectivity of traditional science, instead insisted on holism and intuition. They were concerned with the whole person, including feelings, and with relating the whole person to the natural environment. They were also reacting, in large measure, against the predominance of Jews in medicine, the sciences, and academia.

---

[5]In a similar way, both Dorothy Winsor and Paul Dombrowski have grappled with the question of post facto reasoning in relation to technical communication and the *Challenger* disaster. It is all too easy to assume that because a terrible disaster occurred, someone did something wrong.

This pointedly nontraditional view of science and medicine permeated the Nazi regime and the higher elements of the medical profession, Harrington and others contend. These people feared that mechanistic science held a "nihilistic message of 'technique' and instrumentalism over soul and integrity," which led to the decline of traditional values, the degradation of humanity in general, and the disintegration of society. The opposite, alternative view of what science, especially medical science, should be insisted on naturalistic and holistic investigations and emphasized herbal and homeopathic treatments. More important for our purposes, it insisted on the deliberate assertion of the primacy of spiritual, moral, aesthetic—and political—values over the mechanistic reductivism and objectivism of traditional science. Thus the regime could assert the *primacy* of racial purity and "Aryan" superiority over the disinterestedness of traditional science. Traditional science, on the other hand, assumes everyone to be equal and treats them as individuals within the same human race. This new vision— practically an antiscience—also asserted a salvation mythology. This new, holistic "science," rather than yielding only arid, discrete facts, would yield instead a new world order in which mankind would be purer, stronger, better, and happier.

The products of traditional scientific medicine, Harrington says, were seen as abominations to nature in reversing natural tendencies and hierarchies. The sick should die, and the strong and healthy should prevail and flourish, according to this perspective, and it was the duty of the new "science" to make the strong, healthy, and dominant even more so.

> In this rhetoric, the infirm and disabled—somewhat like the Jews—were *themselves* transformed into metaphors of mechanism. . . . "Machine people" was the disdainful term used by Nazi physician Karl Kotschau—a reference to the fact that such individuals owed their survival to medical technology, and were incapable of surviving outside the shelter of an artificial environment (192).

This primitivism specifically opposed theorizing and abstract reasoning. Traditional science, of course, aimed at developing more sophisticated theories of greater, more general scope from analytical facts. Intuitive reasoning, on the other hand, was held to be more natural, truer, and more authentic because it meshed with the feelings and attitudes of real people. No wonder then that the very best German traditional scientists of the time, from physicists to biologists, fled the country or were exterminated by the regime. Thus, to Harrington, attributing Nazi crimes to the nature of science itself is a serious mistake.

Research in the United States

Though generally in this book ethics has been kept distinct from the law, the Nazi "scientific" information is one ethical dilemma in which legal principles have direct bearing. In our legal system, the means by which information— knowledge—in the form of evidence is obtained can greatly affect its validity and usability. Regardless of the indications of guilt that such evidence might reveal, if it was obtained illegally, then the information is considered not to exist and cannot be used as evidence. Even more stringently, in the case of Miranda rights, without the proper awareness of one's rights so as to allow informed consent to obtain the evidence, the evidence is deemed inadmissable—it in effect does not exist, specifically regardless of whether it would be useful to the prosecution or not. This inadmissability, we should keep in mind, is not just a hope or wish but is a solid legal principle with very real effect. It can determine whether a defendant lives or dies or whether society will be put at risk by releasing from prison an improperly convicted suspect.

This inadmissability principle should be analogous to the Nazi "medical research" dilemma. The subjects of the research certainly did not provide informed consent without coercion or even any consent at all. Even if consent had been obtained, the means or conditions under which the information was obtained would taint the information so strongly and clearly that it would be deemed inadmissable were it to be used in a criminal proceeding. Therefore it should be deemed inadmissable in the scientific community as well, we can reasonably suppose.

Or should it? Some scientists hold their enterprise to be elevated above mundane political considerations that are subject to shifting public opinion. Scientific knowledge is valuable in its own right for its own sake, they say, with a worthiness that transcends squabbles over what is right or wrong, good or bad. After all, the Nazi "researchers" themselves justified their activities as the pursuit of knowledge for its own sake, transcending the niceties of legality, much less ethicality. But is this always a valid line of reasoning? Does science somehow rise above ethical scruples?

We need not limit ourselves to the Nazi regime for examples of means tainting knowledge. The infamous Tuskegee syphilis experiment in the 1920s through the 1940s provide an example from the United States. African–American patients diagnosed with syphilis were prescribed a treatment regimen. In a number of cases, the patients were given only placebos without any effectiveness whatsoever rather than truly effective drugs, entirely without their knowledge and consent, even to the point of outright lying to them. The patients were in effect left untreated but told they were being treated as effectively as possible. The purpose of this deception was to allow the doctors to conduct scientifically a detailed, long-term naturalistic study of the unimpeded progress of the disease. Many of course died of the disease, and all suffered terribly and needlessly.

It is no coincidence that the patients were African–Americans. Such gross mistreatment and deception by white doctors would not have been tolerated on white Americans. As in the Nazi "research," the social status of the subjects and the power relations between them and the "researchers" was in fact essential to the activities themselves. The research would not have been permitted on those of a freer and more equal status.

A more recent episode in U.S. history provides another example of scientific knowledge that was obtained unethically. One of the landmark initiatives of the Clinton administration has been actively to search out and publicly acknowledge unethical scientific research by the federal government. In recent years we have learned of research into the effects of radiation on humans using unwitting human subjects. Some of this was done rigorously enough to be considered valid from the standpoint of scientific methodology, and some not. From the perspective of the Clinton administration and certainly of the public, the point of these revelations is not the scientific validity of the information at all but the ethicality of the method under which it was obtained (and the possible legal and financial liabilities involved). There do not appear to be any claims, however, that this information should be banned or suppressed due to the taint of unethicalness, as there was for the Nazi information. (In the Discussion and Paper Topics section at the end of this chapter, several items ask that you explore further the findings of the investigations by both the Advisory Committee on Human Radiation Experiments and the U.S. Government Accounting Office into the abuse of human research subjects and the misrepresentation of technical information in contemporary scientific research.)

Scientific knowledge can be tainted by the means by which it was obtained even when human suffering is not involved. In the last few decades, increasing concern has arisen about the suffering of animals used in scientific research not only in the field of medicine but also in cosmetics. Manufacturers of cosmetics nowadays commonly advertise prominently when they have avoided using animals to test the safety of their products. Presumably this proclamation resonates with the ethical values of many consumers and influences their decision to purchase the product or not.

As far as ordinary technical communication is concerned, it is doubtful that many technical communicators will ever be involved in such drastic ethical dilemmas as the Nazi and Tuskegee "research" projects. However, some of us will likely be involved in situations in which the means and ends could ethically taint technical information we are dealing with. This taint can be so strong as to warrant considering how, and even whether, the information should be communicated.

In appraising ethical situations such as these, Kant is relevant in a surprising way. One of the most damning indictments of the Nazi documents from Kant's perspective has to do with the scope of these examples. Kant based his theory on close reasoning about human nature and responsibilities. The rational nature of his theory was so fundamental that he states that the categorical imperative of ethics applies not only to all humans but also to any and all "sentient" beings, that is, all those capable of reasoned thought. Even those beings who are not human but are still capable of reasoning should be able to understand the imperative and be guided by it as fully and equally as any human would. To Kant's mind, then, sentience or the capacity for reasoned thinking was even more essential than bare humanness in deciding ethical responsibility. Therefore, even if one were to question whether another person is a full member of the human race by virtue of supposed "Aryan" racial heritage, the simple fact that that person is capable of reasoning would outweigh any argument based on racial identity. Jews—and all the others Hitler sought to annihilate, including Slavs, homosexuals, gypsies, and political dissidents—should have been treated as ethical peers by virtue of their minds.

NAZI TECHNICAL MEMORANDUM

This example of ethical issues involving technical documents from the Nazi regime consists of a single technical memorandum. This infamous document appears on a number of Holocaust Web sites and is the focus of a widely cited article by Steven Katz on the values underlying technological advancement. Though on the face of it the document seems only to be about certain mechanical modifications to motor vehicles, its true subject is something very different.

Katz points out in his notable article that our modern society is heavily influenced by science and technology. This influence is so widespread and strong that technological values such as expediency and efficiency have come to completely dominate our society and reshape its entire value system. The same has happened throughout the technologically advanced countries of the world. As a result, technology, instead of being only a means to pursue worthwhile social goals, becomes a goal or end in itself. It takes on an importance that is desirable in its own right. The Industrial Revolution illustrates well this elevation of technology to become valued and pursued for its own sake. The automobile and its role in the social and economic fabric of our country is only one example. One might say that our society has been shaped by the automobile.

Katz explores in the writing of Adolf Hitler and the Nazi regime that he commanded the complex historical connections among technology, social interests, and metaphysical or spiritual values, all of which led to a shift between means and ends. He shows that technical expediency sometimes can be taken as desirable in its own right, so much so that it can mask the goals or ends for which a technology is being used. This article is especially important because of the vital importance that the Holocaust never be forgotten. A generation brought up on the familiar video presence of Sergeant Schultz and Colonel Klink of *Hogan's Heroes* can all too easily forget the horrifying inhumanity of what went on in the Nazi camps.

Reproduced here is the memorandum (Figure 4.1) written by Willy Just to Lieutenant Colonel Walter Rauff. The place is Chelmno, one of the Nazi death camps. The time is June 1942, only five months after the notorious Wannsee Conference of the Nazis, which outlined the brutal treatment of Jews and the "final solution" of mass murder. Before we examine the document in detail, it

Berlin, 5 June 1942

TOP SECRET!

To: Gruppenleiter II D
          SS-Obersturmbannführer Rauff

Conc.: Technical adjustments to special vans at present in service and to those that are in production.

[1]   *Since December 1941, ninety-seven thousand have been processed, using three vans, without any defects showing up in the vehicles.* Previous experience has shown that the following adjustments would be useful.

[2]   The normal capacity of the vans is nine to ten per square meter. It would appear that a reduction in the cargo area is necessary. This can be achieved by shortening the compartment by about one meter. The problem cannot be solved by merely reducing the number of subjects treated, as has been done so far. For in this case a longer running time is required, as the empty space also needs to be filled with CO [carbon monoxide]. On the contrary, were the cargo area smaller, but fully occupied, the operation would take considerably less time, because there would be no empty space.

[3]   The manufacturer pointed out during discussions that a reduction in the volume of the cargo compartment would result in an inconvenient displacement of the cargo toward the front. There would then be a risk of overloading the axle. In fact, there is a natural compensation in the distribution of the weight. When [the van is] in operation, the load, in its effort to reach the rear door, places itself for the most part at the rear. For this reason the front axle is not overloaded.

[4]   To facilitate the cleaning of the vehicle, an opening will be made in the floor to allow for drainage. It will be . . . fitted with an elbow siphon that will allow for the drainage of thin liquids. Thicker dirt can be removed through the large drainage hole when the vehicle is cleaned. The floor of the vehicle can be tipped slightly. In this way all the liquids can be made to flow toward the center and be prevented from entering the pipes.

[5]   Greater protection is needed for the lighting system. The grille should cover the lamps high enough up to make it impossible to break the bulbs. It

(continued)

FIGURE 4.1 Technical Memorandum by Just on Gassing Vans This figure presents an excerpted version of a technical memorandum by Willy Just to SS Lieutenant Colonel Walter Rauff, in charge of motorized equipment for the SS. It uses masked language, impersonal language, ellipses, and an absolutely technical perspective to disguise its true subject and purpose. Though the memorandum appears to be only about technical changes to a motor vehicle, its true but deliberately unstated subject is Nazi racial extermination policies and the mass murder of people. Numbers in square brackets for paragraphs have been added to facilitate analysis. Reprinted from *Nazi Mass Murder* by Eugen Kogon et al., pp. 228–30.

seems that these lamps are hardly ever turned on, so the users have suggested that they could be done away with. *Experience shows, however, that when the back door is closed and it gets dark inside, the load pushed hard against the door.* The reason for this is that when it becomes dark inside the load rushes toward what little light remains. This hampers the locking of the door. It has also been noticed that the noise provoked by the locking of the door is linked to the fear aroused by the darkness. It is therefore expedient to keep the lights on before the operation and during the first few minutes of its duration. Lighting is also useful for night work and for the cleaning of the interior of the van.

Signed: Just

FIGURE 4.1   Continued

would be best that you read the memorandum yourself in full, keeping in mind its context. As you read, ask yourself these questions:

- When do you first realize what the subject of the document is?
- How is language used in this document to achieve its aims?
- What are some of the ethical dimensions of this language use? Keeping in mind the principles of technical documentation, is this a well-written technical document?

Notice how values are powerfully operative in this document, even though the document and its writer try hard to be neutral, objective, and dispassionate and so try to transcend moral issues.

Let us examine this document from the smallest scale to the largest. At the smallest scale, that of words and sentences, we can say a good deal about this document from the viewpoint of ethics and values. For many people, the first suggestion that something strange is going on here is the glaring absence of key words (the technical term for this is "ellipsis," words and thoughts evoked but not expressed). In the first sentence of the first paragraph, there is no noun stated after the modifying phrase "ninety-seven thousand." Ninety-seven thousand *what*, we wonder? The answer, of course, is people, specifically Jewish people. The writer avoids talking about people probably because that would call up emotional feelings and so detract from the dispassionate stance he is striving for. Instead he talks only about the vehicle.

In the second paragraph the writer continues to avoid talking about people by using impersonal, technical language. This is a clear example of masked language. In the first sentence people are referred to as the "capacity." In the second sentence the cargo area is the focus of attention rather than the cargo itself. In the third sentence the "subjects" are referred to, but there is still no direct reference to them as people. In the sixth sentence the "cargo area" is again referred to but, as before, not the people occupying and defining this space. The writer aims to reduce empty and supposedly wasted space, exactly the opposite concern of the people involved, of course.

In the third paragraph, the "reduction of volume" is also the focus rather than the impact on the people involved. In the fourth sentence, "the load" places itself at the rear, redistributing the load. The writer does not mention why the load does this, but we know, of course, that the people who are the load are terrified. Notice the writer's careful—almost caring—concern for the mechanical vehicle itself, which is heightened by juxtaposition with the total lack of caring concern for the people that the vehicle carries.

In paragraph four, technical details are recommended to facilitate the cleaning of the vehicle. Again the vehicle is the sole focus of the writer as though he has put on blinders to shut out anything but the vehicle. The writer expresses concern for cleaning the vehicle of "liquids" and "dirt" but meticulously avoids mentioning the people who are the source of these eliminated materials and the terror behind it.

In the fifth paragraph, concern for the mechanics of the vehicle begins the paragraph, greater concern than is shown the cargo, of course. In sentences four and five, we learn that "the load" pushes hard against the door as the doors are closed as darkness comes. This "load" is obviously active and alive in the ninth sentence but does not react the way nocturnal animals do to darkness. Sentence seven gets closest to alluding to the humanity of the cargo when "the noise" and "the fear" are mentioned. In this case, though, the writer can refer only to the responses generated by the people, the artifacts of their existence, but cannot bring himself to mention the people themselves. Clearly, throughout this technical document the writer deliberately avoided talking about people by using impersonal language and displaced references and by the simple absence of direct words.

Let us now look at the document as a whole. What is its subject? The subject line of the memorandum refers to "changes for special vehicles," so it appears that the subject is only vehicles. The real, truer subject, however, and the ultimate goal of these proposed changes is the use to which these vehicles are put, namely murdering people.

Note, too, that the subject line not only avoids stating the true subject but also uses impersonal language rather than personal. Apparently the writer finds it easier and more technically effective to talk about things and cool mechanical facts rather than about people killing people, powerful emotions, or ethical burdens. This is true not only for the subject line but also for the opening and closing paragraphs, which work like a picture frame to set and underscore the tone of the entire document. The more graphic and emotionally evocative paragraphs occur in the middle sections, within and controlled by this framework.

Note, furthermore, that the document is "technically" excellent. It has the correct form of a memorandum: a clear, brief statement of purpose at the beginning; a problem plus solution structure; clear paragraphs with a single topic stated at the beginning, then explained and supported by detailed evidence; and a brief, effective closing paragraph. The technical excellence of the composition of the document echoes and tacitly affirms the purely technical excellence that the document aims to accomplish. Therefore, considering only its technique or technical merit, this is a good technical document.

Or is it? Are we justified in evaluating this document *only* in a mechanical, superficial way? Or are we ethically obligated also to consider the context from which this technical information sprang and the uses to which it will be put? If we answer, Yes, we must consider the uses to which this will be put, then should we not also feel obligated to raise the same question for any and all other technical documentation we encounter? Would this consideration apply to our own technical documentation tasks in American industry nowadays, too, for example? The U.S. tobacco industry, in its now-discredited "research" reports rebutting claims that tobacco use causes cancer, provides a ready illustration of how this ethical dilemma can arise in our own society and work today.

We now are only distant, unintended readers of this Nazi document. How do you think the writer and intended readers would have responded to the preceding questions? How would they have responded to our ethical criticism? One response, actually given at the Nuremberg trials, was that the Jews killed were not really people.

To clearly illustrate how the Nazis treated Jews impersonally as things that were not really human, consider the following excerpts from another memorandum. This May 1942 memorandum from Dr. August Becker, a Nazi doctor, to his commander addresses the killing vans as a technology to be perfected. These vans used various gases to kill those trapped within them. Someone had to unload these vans, of course, but it should not be prisoners because then they could not be herded into the vans for their own executions, or they would try more desperately to escape. (What precedes this passage is a discussion in the memorandum about disguising the "death vans" with fake windows and curtains to make them appear to be house trailers.)

> I should like to take this opportunity to bring the following to your attention: several commands have had the unloading after the application of gas done by their own men. I brought to the attention of the commanders of those S.K. concerned *the immense psychological injuries and damages to their health* [emphasis mine] which that work can have for those men, even if not immediately, at least later on. The men complained to me about headaches which appeared after each unloading. . . . To protect the men from those damages, I request orders be issued accordingly (354–355).

Unmarried soldiers were being chosen for this duty because those married with children suffered the worst of these psychological effects. Clearly here soldiers as persons were treated with great importance and caring concern, whereas those executed were treated as things that were not real persons.

Returning to the Just memorandum, we need to consider the social, political, and cultural context in which this document appeared. As Katz explains quite clearly, part of the cultural context was a very strong desire for technical excellence, for doing anything as well as it could possibly be done. The technical values of excellence, effectiveness, efficiency, and expediency came to replace many of the traditional social values of Germany. Traditionally, social goals were discussed in political deliberations, and only afterward were the means for achieving these goals sought. In Nazi Germany, however, the distinction between means and ends became so blurred that what was technically possible came to be sought almost for its own sake. The result is a circular sort of self-justification: What we can do, we should do, largely because we can do it.

Although to our minds this reasoning does not make logical sense, we should nonetheless notice the potential persuasive power of its message. The circular justification of something by itself does not call upon any justification outside itself, of course, but neither does it admit the legitimacy of any criticism outside itself. Thus it cannot be readily attacked either from the outside or the inside within its own framework of values and logic. It is a closed loop and truly close minded.

Many ethical critics of technology contend that modern high technology seems to be pursued for its own sake, as though technical excellence is intrinsically desirable. Langdon Winner, for instance, critiques "autonomous technology" as the feeling that what we *can* do, we *should* do, a sort of imperative that cannot be opposed or defused. Critics such as Winner point out that short-sighted self-justification of this sort was highly operative in the nuclear power industry throughout the world during the 1950s and 1960s and in the chemical industry of the 1950s. Even nowadays in the genetic engineering industry, a rather extremist researcher recently contended with chilling earnestness that cloning of humans would be done, and soon, *because* it can be done.

Just's memorandum presents a real example of technical objectivity taken to an extreme. Though it appears to deal only with impersonal technical details of the efficient operation of a device, its real but disguised subject was of the most intensely personal nature. And its significance is entirely ethical.

To give a fuller picture of how technology and science were conducted by the Nazi regime and a better picture of the overall context of technology and science in the service of racism and anti-Semitism, we will briefly discuss two other technical communications. Besides filling in the broader picture, the primary purpose of the examination of these documents is to see how heavily invested they were with the values of racism and anti-Semitism and to see the rhetorical rationalization of atrocities supported by technological and scientific mentality. They show that the Just memo was not an isolated example.

Figure 4.2 shows a report by Prof. Hirt of the University of Strasbourg about the need for certain Jewish specimens and a plan for meeting that need. The overarching purpose of this plan to acquire skulls of Jews, stated in the first

The Ahnenerbe

The Reich Business Manager (Reichsgeschaeftsfuerhrer)

Berlin, 9 February [19]42

To: SS Sturmbannfuehrer Dr. Brandt

Subject: Securing skulls of Jewish-Bolshevik Commissars for the purpose of scientific research at the Strassburg Reich University

There exists extensive collections of skulls of almost all races and peoples. Of the Jewish race, however, only so very few specimens of skulls stand at the disposal of science that a study of them does not permit precise conclusions. The war in the East now presents us with the opportunity to remedy this shortage. By procuring skulls of the Jewish Bolshevik Commissars, who personify a repulsive, yet characteristic subhumanity, we have the opportunity of obtaining tangible, scientific evidence.

The actual obtaining and collecting of these skulls without difficulty can be accomplished by a directive issued to the Wehrmacht in the future to immediately turn over alive all Jewish Bolshevik Commissars to the field M.P. The field M.P. in turn is to be issued special directives to continually inform a certain office of the number and place of detention of these captured Jews and to guard them well until the arrival of a special deputy.

The special deputy, commissioned with the collection of the material . . . is to take a prescribed series of photographs and anthropological measurements, and is to ascertain, in so far as possible, the origin, date of birth, and other personal data of the prisoner. Following the subsequently induced death of the Jew, whose head must not be damaged, he will separate the head from the torso and will forward it to it's [sic] point of destination in a preservative fluid within a well-sealed tin container especially made for this purpose. On this [sic] basis of the photos, the measurements, and other data on the head, and, finally, the skull itself, the comparative anatomical research, research on race membership, the pathological features of the skull form, the form and size of the brain and many other things can begin.

In accordance with its scope and tasks, the new Strassburg Reich University would be the most appropriate place for the collection of and research upon these skulls thus acquired.

FIGURE 4.2 Report of Prof. Hirt on Securing Jewish Specimens This is the report of Prof. Hirt of the University of Strasbourg of plans for securing skulls of Jews for racial study purposes. The report was submitted to Dr. Rudolf Brandt by a letter of transmittal from SS officer Wolfram Sievers, due to Prof. Hirt's current illness. Prof. Hirt is also noted for acquiring human bodies for anatomical studies, killed on demand, and for "intravital microscopy," the microscopic internal study of living humans. Nuremberg Document Number NO-085.

paragraph, is explicitly and specifically scientific. The university already has collections of skulls from other races and nationalities but few of Jews. (It does not characterize how the skulls of others were obtained, but because they were of non-Jews, they were likely acquired through conventional legitimate means.) The "precise conclusions" sought can only be as a means to categorize people as Jews regardless of any claims to the contrary or to show the inferiority of Jews ("a repulsive, yet characteristic subhumanity") in relation to other races and people. This constitutes the pressing need for "tangible, scientific evidence," as though science is driving Hirt's scheme. The supposed scientific legitimacy of this plan is reflected in Hirt's stating that the new Reich University at Strasbourg—a major academic institution—is the most appropriate place for this collection as though as a point of honor.

The war in the East with Russia makes possible, to his mind, the acquisition of skulls not just of Jews but also of communist Jews, Bolsheviks. Furthermore, the capturing of live specimens allows the researcher to obtain information such as medical history and date of birth that would be difficult to obtain from dead specimens. It is almost exactly like a naturalist doing fieldwork, gathering biological specimens to be "euthanized" and then permanently mounted for close study. The most chilling phrase of the document is "induced death," a scientized evasion in words of the reality: cold-blooded murder. Actually this is a technology, too, a technology of fieldwork for collecting, documenting, and preserving specimens, the purpose of which is scientific, namely, establishing the characteristics of race membership. These people were murdered for the sake of science, supposedly. But the true ultimate purpose was not really scientific at all but political because it aims at the annihilation of the race and all its members. So racial persecution and murder are masked by scientific and technological pretense that affords a supposed rationale. This also enables a radical distancing of the researcher from the humanity of the subjects. The technical document plainly treats people as objects, just as the researchers did.

Figure 4.3 is of a more technological nature although backed up with supposedly scientific studies. The passive voice of the first two sentences is typical of scientific reports and clearly suppresses rhetorical agency, that is, any indication of the person who is doing the experiments. This suppression of agency is not in itself inappropriate for scientific reports. Indeed, because the bedrock of the scientific method is replicability of experiments, a truly scientific experiment must be indifferent to the person performing the experiment: Anyone performing exactly the same procedure should get exactly the same result. In this case, however, we should recognize that this suppression of agency can—and in this case does—have ethical ramifications. If the person is out of the picture, so is the ethical responsibility attached to particular persons.

The first sentences also establish a flat, unemotional tone that simply reports the facts. This tone in itself is not necessarily inappropriate, but, again, it does have ethical ramifications. In this case, the unemotional, matter-of-fact tone is important for what it leaves out, namely, the uses of this scientific information.

TRANSLATION OF DOCUMENT No. NO-203
OFFICE OF THE U.S. CHIEF COUNSEL

Report on Experiments Concerning X-ray Castration

The experiments in this field are concluded. The following result can be considered as established and adequately based on scientific research.

If any persons are to be sterilized immediately, this result can only be attained by X-rays in a dosage high enough to produce castration with all its consequences, since high X-ray dosages destroy the internal secretion of the ovary, or of the testicles respectively. Lower dosages would only temporarily paralyze the procreative capacity. The consequences in question are for example the disappearance of menstruation, climacteric phenomena, changes in capillary growth, modification of metabolism, etc. In any case, attention must be drawn to these disadvantages.

The actual dosage can be given in various ways, and the irradiation can take place quite imperceptibly. The necessary local dosage for men is 500-600 r, for women 300-350 r. In general, an irradiation period of 2 minutes for men, 3 minutes for women, with the highest voltage, a thin filter and at a short distance, ought to be sufficient. There is, however, a disadvantage that has to be put up with: as it is impossible unnoticeably to cover the rest of the body with lead, the other tissues of the body will be injured, and radiologic malaise, the so-called "Roentgenkater", will ensue. If the X-ray intensity is too high, those parts of the skin which the rays have reached will exhibit symptoms of burns—varying in severity in the individual cases—in the course of the following days or weeks.

One practical way of proceeding would be, for instance, to let the persons to be treated approach a counter, where they could be asked to answer some questions or to fill in forms, which would take them 2 or 3 minutes. The official sitting behind the counter could operate the installation in such a way as to turn a switch which would activate the two values simultaneously (since the irradiation has to operate from both sides). With a two-valve installation about 150-200 persons could then be sterilized per day, and therefore, with 20 such installations as many as 3000–4000 persons per day. In my estimation a larger daily number could not in any case be sent away for this purpose. As to the expenses for such a two-valve system, I can only give a rough estimate of approximately 20,000-30,000 RM. Additionally, however, there would be the cost of the construction of a new building, because adequately extensive protective installations would have to be provided for the officials on duty.

FIGURE 4.3   Report on X-rays for Castration This is the report of Viktor Brack, chief administrative officer in the Fuehrer chancellery, on a plan for mass, unwitting castration using X rays, to the Reichsfuehrer-SS and Chief of the German Police. It was part of the program of racial hygiene intended to prevent the reproduction of various people considered undesirable, mostly Jews. Nuremberg Document Number NO-203.

> In summary, it may be said that, regarding the present state of radiological technique and research, mass sterilization by means of X-rays can be carried out without difficulty. However, it seems to be impossible to do this in such a way that the persons concerned do not sooner or later realize with certainty that they have been sterilized or castrated by X-rays.
>
> Signed: BRACK

FIGURE 4.3   Continued

Also omitted are compassion for the people being castrated and sterilized, awareness of shared humanity with these people, ethical concern for the rights of these people, and ethical responsibility for collusion in the entire political policy of racial extermination that this science and technology are supporting. Obviously, because this procedure is done without the subjects being aware of what is going on and even being deliberately deceived about it, there is no ethical concern for the rights of these people.

After explaining the consequences of excessively high dosages and calibrating the dosage for effectiveness with minimal unwanted side-effects, the writer proceeds to describe the procedures for applying this knowledge practically. The document concludes by affirming that the aims driving this research and technological innovation can indeed be achieved. This is exactly the line of development we usually find in any good scientific report: establishing purpose, showing methodology, reporting results, and explaining applications that reflect on the purposes driving the research. So, in a mechanical sense, it works.

For our purposes, however, we should note the extreme emotional and ethical distance separating the researcher from the subject. This allows practically anything to be done to the subject for the sake of the science (though actually for the sake of the politics). The science is much more important than—it is valued above—any rights the subjects might have. We should also note that the report deals explicitly with only the research and the information itself. What it avoids is what is also the topic of this chapter—ethical concern for how information was obtained and how it will be used. With these concerns out of the picture, the researcher probably feels free to focus on the research and such questions as whether an effective dosage for males is 400r or 500r. The same holds for the receiver and user of this information, who can put it to use in a technically effective way while sidestepping any ethical implications.

GRAPHICAL IMAGES

Another way that underlying values found expression in technical and scientific information and documents was through Nazi race laws. These laws, such as the infamous Nuremberg Laws, aimed at ensuring the racial purity of the "Aryan" race by declaring as illegal certain blood relationships of "Aryans" with those of other races but especially Jews. Part of the supposed rationale behind these anti-Semitic laws that institutionalized hatred of Jews was a science of heredity that asserted the superiority of one race over all others. For the Nazis, the "Aryan" race was considered the only truly human race, while other races were less than fully human. Slavs, for instance, were considered to be subhuman but suitable for use as slaves because they did not threaten the racial purity of the "Aryan" line—at least at first. As time went on, though, millions of non-Jews of Slavic and other races were put to death because of their race, too. Jews, on the other hand, were considered right from the start to be not just other-than-"Aryan" but an active threat and a virulent potential pollutant. Therefore any intermingling of Jewish and "Aryan" blood was to be avoided at all costs. This gave rise to bizarre legal contortions such as an "Aryan" secret policeman having to go to court to determine whether he could continue in the police after having received a transfusion of just one pint of Jewish blood after an accident.

The supposed rationalization of this racial hatred was based on the sciences of heredity and evolution. When Darwin delivered his earth-shaking theory of evolution in the mid-nineteenth century, many were quick to apply this new thinking to the human species as well. This thinking, known as social Darwinism, explained human societies in the same evolutionary terms that originally applied only to the animal world. Applying the notion of "survival of the fittest" to humans, for example, implied that some forms of humankind are fitter than others, whereas others are less fit and would naturally die out. The fittest race was naturally superior to and should dominate lesser races. This "natural law" approach to social problems also argued that many of society's problems were the result of highly unnatural practices. In this new scientific vision of humankind, it was assumed that the role of science was to correct unnatural social practices.

One of the ethical problems with this new vision was that, though it appeared to place science above all other considerations, in reality political values came first. Thus science was made to serve politics. Tragically, this new vision took only a selective, distorted view of science as the only legitimate one. All social ills were considered the result of unnatural practices that contaminated bloodlines that were supposed to be naturally pure. Even blood types (A, B, and O) were assumed to have originally been racially specific. The goal of science, especially medical science, was to perfect the "Aryan" gene pool through a program known as eugenics. Thus, for instance, the euthanasia program discussed earlier sought to eliminate "life unworthy of life" that could not survive on its own without unnatural support from others, such as the mentally retarded or the deformed. Later this was expanded to those with lesser impairments, then to those undesirable such as the chronically unemployed or homosexuals, and then in its most virulent form to Jews.

There were many difficulties in forcing pseudoscientific thinking into this political mold. It assumed that a politically congenial sort of science was the only true science, putting politics and racism *before* disinterested science. It also assumed that there are indeed real human races rather than just a single, all-encompassing human race. (A good deal of recent thinking critiques the very idea that there really are many human races as only a contrived notion driven by political interests, such as nationalism and the desire to expand and dominate others.) It also assumed that "Jew" is a race—rather than a religion or a culture—with a definite bloodline that could be biologically identified. Education and research in biological and medical areas came to be controlled by political leaders who would allow only politically correct theories to be taught and practiced and who would not tolerate opposition.

As a result, technical procedures were established to identify Jews. In this way, seemingly concrete, exact technical information was derived in order to support prior racist assumptions. This lent an air of disinterested technical and scientific credibility to what was really quite unscientific and viciously biased. One example of treating racism as a technology is shown in Figure 4.4.

Figure 4.4 outlines the racial intermixing that would occur as a result of marriage or conception. It is a technical illustration showing how to determine the blood purity of an individual from the "race" of his or her parents and grandparents. The Nuremberg Race Laws of 1935 began a policy of racial segregation that required that one's race be clearly and definitely established. This and other racial laws culminated in Jews being gathered throughout all countries occupied by Germany and sent to forced labor or put to death. Only one of the many difficulties in using this chart was how to determine the status of one's ancestors. Usually this was just supposed on the basis of whether an ancestor was a member of a Jewish religious community or had a typically Jewish name. Thus religion, faith, and culture were made to appear to be biologically determined.

For our focus on ethics, we should primarily notice two things about this graphical image. First, notice how the use of a technology gives rise to a distancing between the technician and the subject. The subject being examined for racial determination is categorized in the same way a laboratory animal might be, very neutrally and without regard to feelings such as compassion or common humanity. Second, notice that the use of this technology allows the technician to remain detached also from the values behind the technology. The technician can think, "I am only following the chart," without thinking about how the findings will be used, what purposes they are serving, or what will happen to the person being examined. These uses, purposes, and results are all determined by values, not by objective measurements.

Figure 4.5 also shows how Nazi racism and anti-Semitism were developed into a technology, this time regarding anatomical features especially of the face rather than lineages of blood and marriage. As the persecution of Jews became increasingly violent, the desire to identify Jews and to distinguish them from non-Jews became increasingly determined. At the same, the desire also increased to legitimate racism and anti-Semitism by making them appear to be scientific and technical. This image shows the facial features of a young German

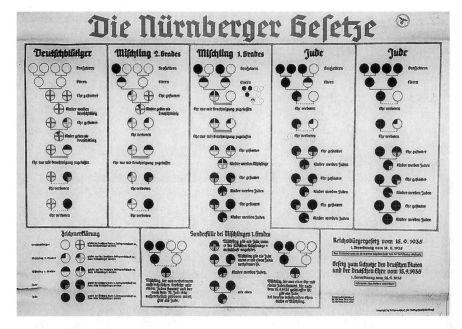

FIGURE 4.4   Chart of the Nuremberg Race Law This chart provides a visual illustration of technical details of the Nuremberg Race Laws in 1935 that forbad or controlled intermarriage of Germans (Aryans) and Jews and defined Jews as noncitizens. Citizenship, of course, established whether one was protected by civil laws. This technical illustration allowed a racial examiner to determine one's racial (specifically Jewish) identity from lineage. It assumed Jewishness to be a racial or genetic matter, reflecting Nazi concerns with racial purity and their adoption of social Darwinism. These racial categories were not really biological categories, though they were represented "scientifically" as though they were. One definition of a Jew, for instance, was "a member of a Jewish religious community." Such technical determinations were literally a matter of life and death.

White space in the circles indicates degree of German/Aryanness. Black indicates Jewishness. Other shading indicates intermixing of identities as "mischling" or half-caste, with varying "degrees." The columns show the outcomes of different parental and grand-parental lineages. The first column results in clear German/Aryan identity, the two rightmost columns result in clear Jewish identity, while the middle columns result in problematic identities of degree with different entailed entitlements.

Photo courtesy of United States Holocaust Memorial Museum Photo Archives

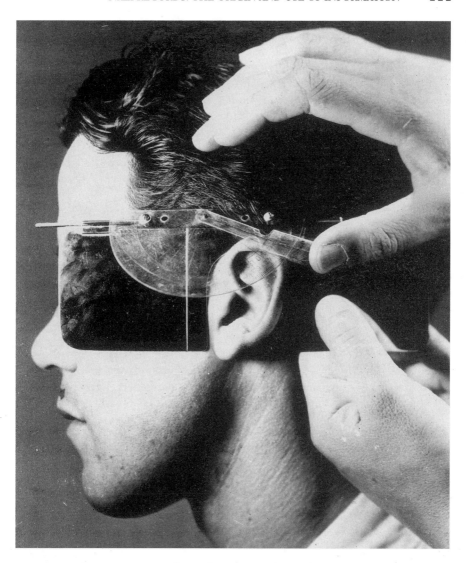

FIGURE 4.5 Measurement of Facial Features of Racial Examination This photo shows the proper technique for measurement of certain facial features as part of a racial examination and determination. It illustrates how the political policy of racism and anti-Semitism was reduced to a technical procedure, making it supposedly objective, impersonal, and incontrovertible. This in turn results in the ethical and emotional distancing of the examiner from the examinee, which is supposedly authorized by the "scientificness" of racial categories. At the same time, it distances the examiner from the ethics underlying the technical policy. In practice, the results of the examination were determined largely by the examiner's intuition and other knowledge because of the poor correlation of features with racial categories.

Photo courtesy of the United States Holocaust Memorial Museum Photo Archives.

man being measured, either to determine whether he is "Aryan" or to build a database of characteristic "racial" features to permit identification of others by race. It illustrates how this technology is put into practice, in this case in a racial examination, so it amounts to a visual technical communication message. Supposedly, facial and other features could be measured to reveal one's true racial identity. These measured indicators were supposed to be truer than other indicators such as marriage records, birth records, and verbal testimony. Technology and numbers do not lie, it was supposed. In reality, though, the measurements were not good indicators and often could not be made to correlate well among themselves or with religious affiliation and could not be done reliably across different investigators.

Notice, as with the racial chart, how the use of a technology gives rise to a distancing between the technician and the subject. Regardless of whatever the technician or subject might think, feel, believe, or want, the measurements give the true story, it was supposed. Notice, again as before, how the technology of making these measurements seems to distance the technician from any ethical responsibility for the values embodied in the technology—the technician is just doing a technical job and doing it as professionally as possible. Last, notice also how the socially (and politically) constructed notion of "race" is reduced to a quantifiable measure and so made to seem technical and scientific—or at least tried to be so reduced. At the same time it tried to reduce Jewishness—whether as a religion, an ethnicity, a culture, a genetic background, or a lineage complicated by marriage—to a matter of technical data and numbers. Even Hitler himself complicated such determinations by explaining that Jewishness was not a race, religion, or culture but practically a unique entity unto itself that defied ordinary classification principles. So what constituted and defined this Jewish "race"? The answer is chilling simple and nontechnical: What Hitler hated.

In practice, the anatomical system of classifying people by race was rather intuitive rather than technical, with the examiner sort of "knowing" beforehand the classification of the person examined. Still later, the technical, quantified techniques were found to be largely unworkable and were generally abandoned. This image, keep in mind, is not just about taking someone's measurements: It was a matter of life or death.

ETHICAL APPRAISAL

The aims and actions of the Nazi regime, as well as the "research" conducted in its name, were abominable. World opinion and legal judgments have singled out these atrocities for lasting attention precisely because they were so terrible, in order that they never be repeated. It therefore seems completely inadequate to call them "unethical," though they certainly were unethical. Our focus, again, is only the particular documents and issues mentioned in this chapter.

In this section we will try to determine how the various ethical theories and perspectives we have discussed would appraise the particular instances discussed in this chapter. We will do the same in the following chapters.

Aristotle.   The Aristotelian perspective based on virtue and cultivated attitudes would ethically condemn the Nazi regime. All the virtues that Aristotle urges are the ones we traditionally associate with goodness and doing the right thing—honesty, integrity, civic responsibility, fairness, and compassion, while nowhere does he recommend pitilessness or violence.

It is true, though, that some critics have expressed serious concern about the pragmatic thrust of Aristotle's ethics and politics, which emphasize expediency, technical excellence, and practical utility. These critics, which include Steven Katz, Dale Sullivan, and Carolyn Miller, are concerned that these aims, can bring about damaging results if they are the chief or only values driving the decisions of social policy-makers.

On the question of whether technical information already on hand should be communicated and used, an Aristotelian perspective would likely urge that it be used. It is available and might benefit others and so should be communicated simply for the sake of that potential benefit. This assessment is consistent with the principle of practical usefulness prominent in Aristotle's ethics, as pointed out by ethical critics.

Kant.   The Kantian perspective assumes the equivalence of all people, indeed of all rational beings. We are all interchangeable for the purpose of deciding what is ethical. We should treat all other people as we would wish they would treat us under the same universal ethical rules. As we have seen, this certainly would include both the Jews and the non-Jews that the Nazi regime deemed as inferior and not fully human. Therefore the persecution and execution of any other people, such as the Nazi regime is noted for, would be entirely unethical.

On the question of whether technical information obtained through Nazi "research" should be broadly communicated, it is unclear how a Kantian would decide. If one would potentially benefit from the communication, then it seems that most people would desire that benefit for themselves. On the other hand, if one were among the victims or a relative of a victim, one might not want the information to be communicated.

Utilitarianism.   Utilitarianism seeks the greatest good for the greatest number, while what is good is defined in terms of usefulness. It is a calculating ethical perspective that balances good against bad, benefits against costs. It assumes that our policy decisions will often benefit some while disadvantaging others. In all the traditional formulations of the utilitarian perspective, there is no reference to anything approaching the utterly vicious harm done to people by the Nazi regime. Nevertheless, it is clear from the documentary evidence that the Nazis saw the good to their "Aryan" race as justifying their putting to death those who were perceived as threatening the racial purity of their race. The calculated good done to the more important group was seen as justifying great harm done to a group considered unimportant. Much more significant than the calculation of good versus harm, however, is the basic assumption made beforehand that some other people are more important and worthy of life than others. No utilitarian theorist supposed such a radical difference in worth among people. Though no one can say for certain whether the great theorists of the utilitarian perspective would ever have approved of the Nazi institutionalization of mass murder, it is highly unlikely.

On the question of the use of the information, a utilitarian would be in favor of communicating and using it. Assuming that the information is already available, then communicating it to others so they can use it could yield only positive benefits and no harm to those now living. With no negatives to oppose the positives, we would have to approve the communication of this technical information.

Feminist and Ethics of Care.   The political dimension is important in a feminist perspective. Authoritarianism in practically any form is criticized throughout the feminist literature. The Nazi regime, of course, is the prime example of authoritarianism, insisting that its orders be followed regardless of personal feelings or preferences. Authoritarianism is even more obvious when we realize that the regime was not just a transient administration such as we have in our American system of government but more a command structure with commands coming from only a single person, Hitler. Key Nazi soldiers, for instance, were required to swear an oath of allegiance to Hitler himself rather than to the government or to the nation. The feminist perspective therefore would find the Nazi regime and its activities to be completely unethical.

Though ethics of care do not stand for all feminist perspectives, they are supported by some feminists and the nontraditional, nonauthoritarian thrust of them resonates with a good deal of feminist thinking. The hallmark of an ethic of care is the caring relationship. Both a caring attitude and the valuing of the relationship for the sake of the relationship itself are key assumptions. Both of these assumptions are violated in the Nazi actions. The Nazis showed an absolutely uncaring attitude toward their victims and absolutely refused to maintain any relationship with them. This is shown in their treating people not as persons but as objects. Their actions were therefore utterly unethical.

On the question of whether the information should be communicated and used, it seems that a caring concern for those now living would require us to communicate and use this information. Not to communicate it could potentially work against those we care for, and so would not be recommended. In addition, the potential usefulness of this information to everyone regardless of race or ethnicity would show an ethically justifiable contempt for the racism of the Nazis, flatly opposing their aims.

CONCLUSION

In this chapter we have seen that, aside from the information itself, both the manner in which information was obtained and the manner in which it will later be put to use have significant ethical implications. As technical communicators, we have a responsibility to be sensitive to these issues as we consider how, or even whether, to communicate such information. This does not mean that the ethical burden lies only with us, of course, instead of with subject matter experts and end users. It only means that the ethical involvement of these others does not entirely relieve us of an ethical burden, too.

We have also seen that the manner in which information is communicated has significant ethical implications, too. This includes not only the language but also the absence of language, the voice, the organization, the purpose, and the values at work behind the scene in a communication. On the one hand, we saw how a radical reversal of meaning is communicated in language, such as in "life unworthy of life" and in the "healing-killing" paradox. On the other hand, we have seen how a "technically" correct memorandum can carry an enormous ethical load even when it is carefully objective and impersonal. In fact, the objectivity, impersonalness, and emotional distance that we often find in technical and scientific investigations can at times be carried to extremes with devastating consequences.

Topics for Papers and Discussion

1. Industrial espionage, spying between industrial organizations, is an increasingly important fact of life in the business world. It is especially prevalent in fiercely competitive fields, such as the software industry, in which any technical advantage can affect the very survival of a company. Suppose that as a technical communicator you are asked to prepare and distribute documentation for a software package that includes an exotic new algorithm. You have learned that this algorithm is based on one that was obtained illegally and that it is rightly the property of another company, Callisto Inc.

You have also learned that your company, Io Industries, is in a precarious financial situation. If a competitor can market a major new software package with substantial advantages over your own, your company could soon go bankrupt. Defusing the potential advantage of Callisto Inc., however, would maintain the solvency of your company and your job. On the other hand, blowing the whistle might cost you your job. Simply refusing the task might have the same result. What are the ethical issues involved here? What factors would influence your decision whether to publish this information or any other action you might take?

Let us change the situation a bit. Suppose that you are a technical communicator for the competing company, Callisto Inc., instead. Would the fact that Io Industries illegally obtained vitally important industrial information make it seem to you ethically unacceptable for any technical communicator at Io Industries to publish and benefit from that information? Further suppose you are an experienced technical communicator who serves on the ethics committee of a major technical communication professional organization. Your ethics committee has been asked to render a judgment as to whether a technical communicator at Io Industries should publish the information. How would you decide and why?

The purpose of this discussion is to see if altering the immediate personal situation and relationship would affect your decision. If it would, does this mean your ethical standards change as your self-interest is affected? Do you think, on the other hand, that ethical standards should be universally binding regardless of your personal involvement? Discuss and explain your ethical opinions in small groups.

2. Obtain copies of the Nuremberg Code of 1946. One possible source is the Web site of the MacLean Center for Clinical Medical Ethics at the University of Chicago (http://ccme-mac4.bsd.uchicago.edu/CCMEPolicies/nuremberg.) Another possible source is *Trials of War Criminals Before the Nuremberg Military Tribunals under Control Council Law No. 10: Nuremberg October 1946-April 1949*, Washington: U.S. GPO (n.d.), Vol. 2, pp. 181–82. Your library can direct you to other sources, too.

This international code for human experimentation grew directly from the shocking revelations of the Nuremberg trials. Analyze and appraise this code from several of the ethical perspectives we have discussed, especially the Kantian and utilitarian ones. How are these perspectives represented, or not, in the code? Do ends or purposes justify means in this document? Is there a utilitarian weighing of the benefit to the many over harm to the few?

For further study, discuss the article " 'The Only Feasible Means': The Pentagon's Ambivalent Relationship with the Nuremberg Code" by Jonathan D. Moreno. Moreno discusses, among other topics, the contention by the U.S. Department of Defense that secret research on the effects of radiation on humans was necessitated by the dire national security issues of the times concerning the effects of possible nuclear warfare. Survey the attitudes of students in your class on Moreno's points.

3. Locate and study another single actual document concerning Nazi medical "research" or technical documentation. Identify any terminology, language use, sentence structure, or organizational pattern that is ethically problematic. Analyze the ethics or values both implicit and explicit in the document, and present your own appraisal of these values.

4. Consider again the EPA's position on Nazi research on the effects of phosgene discussed in this chapter. Suppose instead that United Nations inspectors just recently discovered a cache of technical documents on research that Sadaam Hussein had performed on unwilling political prisoners on the effects of a nerve gas.

The U.S. government has declared that the information was immorally obtained and therefore should not be disseminated to medical scientists in this country, even those who are searching for antidotes to potential chemical and biological warfare agents. You agree with your government's position.

Suppose, next, that the military draft was reactivated in America, and you were about to be sent to the Persian Gulf area on a Desert Storm-like mission. As a result, you stand a fair chance of being exposed to the nerve gas Sadaam Hussein has stockpiled in his arsenal. Would this change your judgment about whether the information should be disseminated to U.S. researchers?

As an alternative, suppose the same scenario as the preceding, with this change. Now you are the only survivor of your Iraq-born family. Only two years earlier, all your immediate family was sentenced to prison for plotting the overthrow of Sadaam Hussein. You had watched as your family were hauled off one by one to serve as guinea pigs for nerve gas research, never to be seen again. You, though, fled the country to the United States. In the business news on television today, you heard that the stock of AntiGen, a new biotechnology firm, had just doubled in price. This jump was the result of AntiGen's having been contracted by the U.S. military to develop an antidote to Sadaam Hussein's nerve agent.

The firm was assured of stellar success and astronomical profits because of the information they had acquired from the Iraqi technical documents through the U.S. government. Some stockholders had instantly become wealthy from the information obtained through the suffering and death of your father, mother, sister, and brother. What is your ethical judgment of the actions of the U.S. government and of AntiGen?

5. Objectivization, the treating of a person as an object to the detriment of that person, is one of the ethical concerns most frequently expressed about technical and scientific research and reports. Among those concerned are feminist critics of science and technology such as Beverly Sauer and Lee Brasseur. Read the article by either of these feminist critics listed in the References section, summarize the article, and offer your own ethical appraisal of it and the issues it raises. Then generalize the principles it presents to other technical communication situations you might find yourself in.

Objectivization, at least in its excessive forms, implies an impersonal attitude and a substantial emotional distance from the person who is the research subject. This excessive psychological, social, and emotional disengagement can be very harmful to that person in many ways. This topic has been investigated and reported in various psychological and sociological journals as well as in communication and rhetoric journals.

A recent article in *Scientific American* considers the opposite dimension of the relationship between technical and scientific experts ("The Placebo Effect," 278.1 [January 1998]: 90–95). It reports, among other findings, that a caring relationship between a doctor and a patient can be beneficial and therapeutic in itself aside from any medical intervention. A relationship of this sort involves a close emotional distance and an affirmation of the patient as a unique individual with personally meaningful concerns, not just as a collection of symptoms. Summarize this article, especially the "healing environment" in contrast to the Nazi "healing-killing paradox." Be sure to discuss the research reports that support its claims. Relate its findings to technical or scientific investigations that you as a technical communicator might encounter.

6. The Hastings Center is a nonprofit institution concerned with investigating and publishing ethical issues in biomedical fields. Read the article "Ethics in the Daily Language of Medical Discourse" by Suzanne Poirier and Daniel J. Brauner. Discuss how objectivization can occur in medical discourse even in situations that are not experimental, such as in daily case reports. Discuss also the authors' suggestions for cultivating a more "human dimension" in medical discourse. Do you agree or disagree? Can you think of additional suggestions of your own?

For further study, read also "From the Ethicist's Point of View: The Literary Nature of Ethical Inquiry" by Tod Chambers. It discusses the important role in language and discourse in shaping our ethical concepts and forming our opinions.

7. Recently the Clinton administration actively investigated and publicly revealed secret tests on the effects of radiation on humans in the decades just after World War II. These tests were done without the knowledge or consent of those being subjected to the radiation, and so were usually unethical. The Advisory Committee on Human Radiation Experiments (ACHRE) has already issued its final report. In addition to radiation experiments, the committee found ethical lapses in other scientific or medical research as well.

For a short paper, read only the Executive Summary (12 pages long) of the Final Report of the committee. Summarize its findings, conclusions, and recommendations, draw connections between it and the ethical discussions in this chapter, and suggest similar situations a technical communicator might encounter today. The report can be located through the U.S. Government Printing Office or a university library or obtained online at either http://raleigh.dis.anl.gov/roadmap/achre/summary.html or http://tis-nt.eh.doe.gov/OHRE/roadmap/achre/summary.html.

For a term paper, read the entire Final Report and write a longer report addressing the same items just listed. Include a discussion of the article "Judging the Past: The Case of the Human Radiation Experiments" by Allen Buchanan discussing the relevance of the ethical appraisal of past events for present and future activities.

8. The U.S. Government Accounting Office (GAO) has conducted an investigation similar to that of ACHRE (see item 7): "Scientific Research: Continued Vigilance Critical to Protecting Human Subjects" (HEHS-96-72, March 8, 1996). It can be accessed online through http://www.gao.gov/AIndexFY96/abstracts/he96072.htm. Another source is the National Security Archive at

George Washington University, http://www.seas.gwu.edu/nsarchive/radiation. Summarize this report as it relates to the topics of this chapter. Suggest situations in which a technical communicator might encounter similar ethical dilemmas. Consider, too, the statement, "No practical level of oversight can guarantee that each researcher will protect subjects with complete integrity" (page 1 of online abstract). How does this relate to the personal ethical burden discussed in this book?

REFERENCES

Advisory Committee on Human Radiation Experiments. *Final Report of the Advisory Committee on Human Radiation Experiments.* Washington, D.C.: GPO 1996. (Related sources are Supplemental Vols. 1, 2, and 2a; Executive Summary and Guide to Final Report; and Interim Report of October 21, 1994.)

Angell, Marcia. "The Nazi Hypothermia Experiments and Unethical Research Today." *The New England Journal of Medicine* Vol. 322 (5/17/90): 1462–64.

Annas, George J., and Michael A. Grodin. *The Nazi Doctors and the Nuremberg Code: Human Rights in Human Experimentation.* New York: Oxford University Press, 1992.

Barondess, Jeremiah A. "Medicine Against Society." *JAMA: Journal of the American Medical Association.* Vol. 276/20 (11/27/96): 165–71.

Becker, August. Letter to SS-Obersturmbannfuehrer Rauff, 16 May 1942. Whitney R. Harris. *Tyranny on Trial: The Evidence of Nuremberg.* Dallas, Texas: Southern Methodist University Press, 1954.

Berger, Robert L. "Nazi Science—the Dachau Hypothermia Experiments." *The New England Journal of Medicine* Vol. 322 (5/17/90): 1435–40.

Brasseur, Lee. "Contesting the Objectivist Paradigm: Gender Issues in the Technical and Professional Communication Curriculum." *IEEE Transactions on Professional Communication* 36.3 (September 1993): 114–23.

Buchanan, Allen. "Judging the Past: The Case of the Human Radiation Experiments." *Hastings Center Report* 26.3 (1996): 25–30.

Caplan, A. L., ed. *When Medicine Went Mad: Bioethics and the Holocaust.* Totowa, New Jersey: Humana Press, 1992.

Chambers, Tod. "From the Ethicist's Point of View: The Literary Nature of Ethical Inquiry." *Hastings Center Report* 26.1 (1996): 25–33.

Dickman, Steven. "U.S. Role Alleged in Cover-Up of Researchers Guilty of War Crimes." *Nature* Vol. 335 (10/6/88): 481.

_____. "Scandal over Nazi Victims' Corpses Rocks Universities." *Nature* Vol. 337 (1/19/89): 195.

Harrington, Anne. "Unmasking Suffering's Masks: Reflections on Old and New Memories of Nazi Medicine." *Daedelus* Vol. 125/1 (Winter 1996): 181–205 (For fuller treatment, see Harrington's *Reenchanted Science: Holism in German Culture from Wilhelm II to Hitler,* Princeton University Press, 1996).

Kater, Michael H. *Doctors under Hitler.* Chapel Hill and London: University of North Carolina Press, 1989.

Katz, Steven B. "The Ethic of Expediency: Classical Rhetoric, Technology, and the Holocaust." *College English* 54.3 (1992): 255–75.

Lifton, Robert Jay. *The Nazi Doctors.* New York: Basic Books, 1986.

Longo, Bernadette. "An Approach for Applying Cultural Study Theory to Technical Writing Research." *Technical Communication Quarterly* 7.1 (Winter 1998).

Moreno, Jonathan D. " 'The Only Feasible Means': The Pentagon's Ambivalent Relationship with the Nuremberg Code." *Hastings Center Report* 26.5 (1996): 11–19.

Nazism: A History in Documents and Eye Witness Accounts. 1941–1945. Vol. 2. Doc. 913. Washington, D.C.: U.S. GPO.

Noakes, J., and G. Pridham, eds. *Nazism, 1919–1945: A History in Documents and Eyewitness Accounts.* Vol. II. New York: Schocken Books, 1988.

Poirier, Suzanne, and Daniel J. Brauner. "Ethics in the Daily Language of Medical Discourse." *Hastings Center Report* 18 (1988): 5–8.

Sauer, Beverly. "Sense and Sensibility in Technical Documentation: How Feminist Interpretation Strategies Can Save Lives in the Nation's Mines." *Journal of Business and Technical Communication* 7 (January 1993): 63–83.

"Scientific Research: Continued Vigilance Critical to Protecting Human Subjects." Government Accounting Office. HEHS-96-72, March 8, 1996. Washington, D.C.: GPO, 1996.

Sun, Marjorie. "EPA Bars Use of Nazi Data." *Science* Vol. 240 (4/1/88): 21.

NB: Re: Holocaust documents. There are many on-line sources for technical documents from the Holocaust, including the following:

- Cybrary of the Holocaust at http://www.remember.org
- The Simon Wiesenthal Center at http://www.wiesenthal.com/bibliog/
- The Forgotten Camps at http://www2.3dresearch.com/~june/vincent/camps/camps.engl.html
- U.S. Holocaust Memorial Museum at http://ushmm.org

CHAPTER 5

CHALLENGER DISASTER
INFORMATION VS. MEANING

The space shuttle *Challenger* exploded before the eyes of millions of television viewers on January 28, 1986. This disaster cost the lives of the seven astronauts and brought the shuttle program to a long halt until its causes could be identified and corrected. It also tarnished the image of the National Aeronautics and Space Administration (NASA) in the eyes of many Americans. Subsequent investigations of the explosion by two federal panels—the presidential commission and the congressional committee—showed that this event was not only a disaster but also a tragedy, an instance of failed judgment.

The investigations reveal a great deal about the nature of technical communications. They demonstrate the critical importance of clear communications in highly technical systems such as the shuttle, the powerful role of complex social forces in shaping communications, and the close interplay between values and language in communications. They show, too, how ethical responsibility can be reflected in highly technical documents. Tragically, they also illustrate how differences in organizational power can negate even the most ethically responsible of technical communications.

In addition to the reports of the investigations, the wealth of interpretations and critical commentaries that arose from the investigations themselves reveal an important aspect of technical communication. They show the role of values and expectations in shaping interpretations even of concrete, factual information. They also show that the ethical assessment of real organizational and technical issues often is very complicated, difficult, and innately problematic. Ethical questions rarely yield easy answers.

In this chapter we ethically examine the events leading up to the disaster, the reports of the two governmental investigations of the disaster, and some of the varied interpretations that have been published about the disaster. We will see that values, whether philosophical or technical, played a powerful role in shaping technical discourse about this disaster. This role is inherent to varying degrees in all technical communications and cannot be evaded. Even the most highly technical documents reflect values and either carry out the ethical responsibilities of the technical communicator or do not. We will see, too, that the audience has a great deal of responsibility in technical communications, sometimes even more than the communicator has. Because this book is directed toward technical communicators rather than their audiences, however, we will focus only on how technical communicators can best carry out their ethical responsibilities in their communications.

When considering the terrible *Challenger* disaster, we need to maintain a realistic perspective. The shuttle system is innately dangerous, as all manned space flights always have been and will continue to be for the foreseeable future. All the shuttle astronauts accept this fact. They also, however, assume that whatever can be done to minimize the danger will have been done. In the case of *Challenger*, unfortunately, this assumption was mistaken.

NASA and its corps of astronauts had experienced other disasters in the past. In the early days of the Apollo program, a fire broke out in the very first Apollo spacecraft (known as Apollo 204) during a training exercise on the ground on January 27, 1967. Even with technicians and equipment present just outside the spacecraft, the fire developed so fiercely that the three astronauts inside died immediately. The primary reason the astronauts did not escape was that the spacecraft used an atmosphere of pure oxygen, which aggressively supports combustion. In addition, the hatch on the spacecraft could not be opened quickly from the inside. Technicians outside had hardly begun the complicated task of unbolting the hatch before the astronauts were dead.

Later, in the case of Apollo 13, as most of us now know from the popular movie of the same name, three astronauts almost lost their lives during a flight to the moon. An electrical short in a tank of oxygen set off an explosion that ripped open the side of the spacecraft, damaged important equipment, and depleted vital oxygen supplies. Only through determination, ingenuity, and plain good luck were the astronauts able to return to earth alive.

In both these earlier cases, the subsequent investigations identified the causes and erroneous decisions. Neither case, however, captured the attention of critics and the public the way the *Challenger* disaster did. The reason for this difference, I believe, is largely that the subsequent investigations revealed not only causes and errors but also a long chain of failures of ethical responsibility. This chain of failures spanned many years, from weaknesses in the early design of the shuttle, through years of dismissals of alarming evidence, to unwise decisions made on the morning of the launch. These were instances not of technical error or insufficient information but of flawed judgment and decision making, many of which are recorded in technical documents.

Focusing on *Challenger*, we first examine the two governmental reports of the disaster, and then examine how conceptualizations and assumptions shaped technical communication in two key developments, and then examine the "smoking gun" memorandum. Tragically, this well-written technical document was rendered ineffective by the unwarranted assumptions of its audience.

TWO GOVERNMENTAL REPORTS

The report of the commission appointed by the president, commonly known as the Rogers Commission, provides important insights into the way that ethics (in the sense of particular values) plays an important role in technical communication. As a written document, it shows difficulties in its proportions, its focus and emphasis, and its language. It also points out difficulties of clarity, purpose, and aims. All of these difficulties, I believe, reflect value considerations, either positively or negatively, that were operative among its writers. The other governmental report, that of the congressional committee, is discussed here for comparison to that of the presidential commission.

Report of Presidential Commission

The presidential report is voluminous, covering five volumes and over 1,700 pages. Most of the material in these volumes is somewhat relevant to understanding the disaster, but not strongly so. The relevance is often weak and only tenuously expressed. Most technical communicators and technical readers expect the proportions of a document to reflect importance; the most space should be devoted to the most important material. The proportions of the presidential commission's report, however, are lopsided, leaving the misleading impression on the reader that the matters covered at substantial length are of substantial importance. Two examples of this misproportion are the sections dealing with the escape system and landing options and with the tire and braking system.

The escape system—actually, the lack thereof—and landing options section covers 11 pages out of the total of about 205 pages of the report per se in Volume I (Vols. II through V are long appendices to Vol. I). In comparison, the section on the "communications" problems, which were the primary immediate cause of the disaster and therefore much more directly relevant, covers about 22 pages. Every one of the investigators understood that escape from the *Challenger* explosion was impossible. No escape system ever considered for the shuttle, whether before or after *Challenger,* has been intended to meet such an extreme contingency. The same holds for the braking system. The five pages on the history of problems with the braking system are largely irrelevant to tracing the causes of the disaster. The explosion and aerodynamic forces instantly ripped apart the shuttle vehicle, leaving nothing intact to fly back to base and to brake for a landing. The weaknesses that were identified in the braking system of all shuttles, therefore, were almost beside the point.

To be sure, the sections of the presidential report dealing with escape and braking are not totally irrelevant. They are relevant, though weakly, to the *Challenger* because they indicate the general atmosphere of unconcern for safety and the possible recovery of the crew in some emergency conditions. They identify weak links in the safety net that should be remedied if the shuttle program were to be continued, which it was.

The point, however, is the proportion. The amount of space given to these two topics, as well as several others, is out of proportion to their overall relevance and importance. If we fail to recognize this disproportion, though, we are left with the erroneous impression that these topics were significant factors in the disaster, although actually they were not. They could instead have been reported in the voluminous appendices. In particular, we are left with the misleading impression that these two sections together were nearly comparable in importance to that of the communications discrepancies, though they actually were not.

What were the values operative behind the scene when these sections of the document were drafted? We do not know for certain what happened or why, but three reasonable possibilities come to mind. First, the disproportion was entirely unconscious and unintended. The commission may simply have wanted to report on anything and everything they explored so that all is on the record. This amounts to writing as a memory dump, which is not a sound approach to any writing project. Furthermore, this does not explain why these sections were not appended rather than included in the body of the report. A careful writer should select from the wealth of data available, highlighting things that are most important and downplaying if not excluding things that are relatively unimportant; this is what the reader expects of the writer, too.

Second, the writers could have chosen to include all the information they collected regardless of the strength of its relevance in order to convince the readers that they had conducted a thorough investigation. The raw amount of information is no guarantee of thoroughness, correctness, or appropriateness, but it does at least suggest these. This also does not, however, explain why these sections were not appended instead of included in the report.

Third, the writers could have included a great amount of information in order to leave the impression that all these factors played significant roles, that none was markedly more significant that any others, and that no causative factor could fairly be singled out as more important than others. This would be the natural, but here mistaken, impression from seeing an imposing mass of information. As in the second possibility, the wealth of information suggests thoroughness and correctness, lending by sheer mass a suggestion that the investigation was thorough and that the conclusions, whatever they might be, are correct. As we will see later, however, in the presidential commission's report the conclusions do not quite follow logically from the mass of evidence and testimony they collected.

Quite possibly this great mass of information was presented partly in order to mislead or at least to complicate rather than clarify. The purpose might have been to divert attention from specific instances of poor judgment or negligence. The supposed effect would be to make the disaster appear to be only an unforeseeable and hence unavoidable accident rather than a tragic failure of judgment and responsibility. In the former case, the deaths would likely be seen as regrettable but largely a fateful consequence of an inherently dangerous job. The public would be left with the image of NASA as grieving but guiltless. In the latter case, however, the deaths would be seen as avoidable and the result of blameful conduct. Thus culpability and liability would become prominent issues, making the sting of these deaths more bitter and more long lasting for the survivors.[1] It would tarnish the image of NASA, demonstrating that they do not always do the right thing for noble reasons.

Differences between Reports

Formal reports were issued by both federal panels investigating the *Challenger* disaster, the presidential commission and the congressional committee. The two reports differ in significant ways. The presidential report differs from the congressional report in its focus, in its conclusions and recommendations, and especially in its unwillingness to assign ethical responsibility.

More than Information.   It is common knowledge that presidential appointees and the congress—representing the executive and legislative branches respectively—often differ. Indeed, differences are routinely expected, for this is a fundamental assumption of the system of checks and balances built into our governmental structure.

These differences in the case of *Challenger* are particularly noteworthy, though, for two reasons. First, the differences center around ethical responsibility, which strongly suggests that this area is a very real crux in understanding the disaster. Second, the congressional and presidential panels considered almost exactly the same body of information—physical evidence, documentary evidence, and testimony—yet came to different conclusions. The congressional committee accepted and adopted practically all the information considered by the presidential commission. (It also, to be sure, went somewhat beyond that information in calling some additional people to testify and in asking some of the same people to testify separately to them, answering their own questions.) What is important for us to notice is that the evidence and testimony were almost identical, yet the conclusions drawn by the two bodies are distinctly different.

These differences suggest two possible conclusions. First, all bodies of information are themselves innately problematic to some degree; that is, any body of complex information in itself does not necessarily determine its own meaning. Instead, meaning comes from the confluence of information with assumptions, interests, goals, and values. Meaning is therefore socially contingent

---

[1]The question of liability actually was raised by some of the surviving families. (Other families were barred from suing for liability because their deceased astronauts were on military duty while on the mission.) The potential suers settled with NASA out of court on the condition that they never attempt to hold NASA liable in court and that they never discuss the matter publicly. So the question will never be answered openly. As the heading of a *New York Times* article put it, "Shuttle Accords Avert a Showdown" (January 2, 1987, p. 3). See references section for additional sources.

and constructed and does not spring fully formed from a body of data. Therefore, different people looking at the same evidence would not necessarily draw the same conclusions.

Second, let us suppose the opposite of what was just stated; let us assume instead that meaning *does* spring fully formed from any body of data. If that is true, then two different meanings cannot be ascribed to the same information without one of them being mistaken or incorrect. In that case, we as readers are left to sort out which is correct and which incorrect. (Another, less likely possibility is that neither is correct.) But because both meanings supposedly came from the same body of information, our judgment about which of the two is correct would have to be guided by some source other than the information itself. Whatever this source might be, it would have to come from the people examining the information and choosing what the information means whether consciously or unconsciously. We, though, as responsible technical communicators should ensure that our own choosing is both conscious and conscientious. In each of these two possibilities, notice that we are compelled to look beyond only the information or evidence itself to other factors such as values, interests, expectations, and presuppositions.

Confusing Language.   The conclusions and recommendations sections of the presidential commission's report, the vital heart of the document, show problems of language and conceptualization. In several key places, it uses different language in discussing what amounts to the same thing. This lack of consistency of usage is confusing, of course. The standard advice in technical communication textbooks is to choose clear and precise terminology and then to use it consistently.

In the concluding section of the commission's report, the commission lays out its understanding of the causes of the disaster (Vol. V, 82). The first paragraph says about the decision to launch that "the decision was flawed." The second paragraph says, somewhat differently, that "the decision-making process was flawed." And the third paragraph says that the cause was "failures in communication." What does all this mean? If the decision itself was flawed, then the line of reasoning was invalid or the conclusion was unwarranted; that is, the conclusion should not have been logically drawn from the available information. If, however, the decision-making process was flawed, then this would seem to focus on all the means by which the information is transmitted to the decision makers and the steps in the process, but not on a mistake of the decision-maker per se. If, still further, the cause was failures in communication, under the traditional transmitter, medium, and receiver model of communication, this would seem to suggest that someone failed to transmit something, or that what was sent was garbled or lost by the medium of transmission, or that the receiver did not accept or understand what was transmitted. These three different expressions represent three different explanations. Did someone fail to do something required by procedures, or did everyone do what was required by an inadequate system of procedures, or was the course of action chosen on the basis of available information actually incorrect and unwarranted? These three reasonable readings of the commission's language about the same focus of causation are all inconsistent among themselves.

Not only is this different language usage inconsistent, but it is also vague and ambiguous. Which of the different representations is most correct, right on the mark? Notice that the very fact that different terms were used cannot help but make the meaning ambiguous. Does it all mean one thing, or another, or a third meaning, or are they all off the mark? This ambiguous language leaves the reader with the task of trying to sort out the correct meaning, a task that should have been undertaken and completed by the writers. All too often, when this difficult sorting out of meaning is left to the reader, it ends up poorly done or undone.

Even more important for our purposes, this different language leaves the reader confused and perhaps misled. What do the writers really mean; what is really going on here; and why? These are all questions that are left unanswered. And remember that the usual reason readers read functional documents such as technical reports is to answer important, practical questions. The writer's role is to strive to answer these questions as fully, fairly, and clearly as possible, which the presidential commission's report apparently did not do.

Conclusions Do Not Follow Logically.   The last point regarding the presidential commission's report is that it fails to address clearly and squarely the question of ethical responsibilities. As a result, the conclusions and recommendations sections of this report amount to a non sequitur (a Latin expression meaning that the conclusions not follow logically from the information beforehand). It does this in three ways. First, the conclusions and recommendations sections does not point a finger of ethical blame toward anyone regarding any particular persons or decisions. The evidence and testimony recorded in the report itself, however, strongly suggest that some ethical responsibilities were not fulfilled. Second, the recommendations section recommends that additional procedures be instituted in order to prevent similar disasters in the future, even though the evidence and testimony clearly indicate that the procedures in place were already adequate. Third, the implicit suggestion left by this call for more, new procedures is that the procedures were at fault or to blame. Of course, strictly speaking, impersonal procedures cannot be "blamed" for anything because blame is a notion associated with conscious and responsible choosing, which can be done only by people. If, though, impersonal procedures are the problem, then no one is to blame. The disaster then amounts only to an unfortunate turn of events.

To explore the comparative roles of people and things, in an article published several years ago, I examine in detail all the contributing causes, events, and decisions leading up to the disaster as presented within the body of the presidential commission's report itself ("The Lessons of the *Challenger* Investigations").

I appraised whether each instance fell into either of two categories. In cases in which an actual person made a conscious choice from a range of possible options that were not already dictated from external considerations, I called that a case of personal responsibility. In cases in which a decision was made but somewhat mechanically, the choice being dictated by procedural directives, I called that a case of procedural decision.

To illustrate this distinction between personal and procedural decision making, consider how the field of education handles possible evidence of the physical abuse of children. If any evidence of possible physical abuse is detected by, say, a teacher, it *must* be reported to the school principal, who *must* in turn report it to the legal authorities. This is a procedural decision or course of action, the direction of which is already determined by external considerations, namely the law. The person really has no alternative (which is just how it should be). Determining what precisely is an instance of physical abuse, however, is often (though certainly not always) less clear and a matter of personal judgment. Repeated instances of broken limbs would be strongly suggestive of abuse, of course. A couple of occasions of bruised thighs would be less clear, however.

Let me offer an illustration from my own experience. A few years ago, my two-year-old daughter cut her forehead on the corner of a kitchen counter one day. I rushed her to the hospital for stitches. I was taken aback, though, when the staff began to ask me not about her cut but about several bruises on both her shins. I explained that she is an especially rambunctious child who runs into things many times a day and is not deterred by this from continuing to run and play with abandon. I was angered at even the suggestion that I had abused her. Another part of me was thankful, though, that, given unclear evidence, the staff erred on the side of protecting the welfare of the child rather than on the side of conveniently choosing not to notice her bruises.

Returning to the presidential commission's report, of the twelve major instances I examined, all of which were highlighted already by the presidential commission itself, I found that most involved personal rather than procedural decision making. Therefore, personal responsibility was a key factor in crucial decisions, as indicated by the evidence and testimony of the report itself. The conclusions and recommendations sections of the commission's report, then, do not follow logically from this information because they do not say anything about personal responsibility and because they urge new procedures, suggesting that procedures were the problem. Overall, I conclude that the presidential commission's conclusions are flawed in evading the issue of personal responsibility.

An authoritative indication that my appraisal is reasonable and warranted is the congressional committee's report. Recall that the congressional committee utilized practically all the same physical and documentary evidence and testimony generated by the presidential commission's investigation. Its conclusions were very different from those of the presidential commission, however.

The congressional report is explicit and clear. It says, for example, that "the failure was not the problem of technical communication, but rather of technical decision making" (30). It is clear too in not "blaming" the procedures but in assigning responsibility to people: "It seems clear that the process [i.e., procedures] cannot compensate for faulty engineering judgment" (70). Its strongest statement is the following:

> [T]he committee finds no basis for concluding that the Flight Readiness Review procedure [which begins more than a month before a launch up through the launch day] is flawed; on the contrary, the procedure appears to be exceptionally thorough and the scope of the issues that are addressed at the FRRs [flight readiness reviews] is sufficient. . . . However, the Flight Readiness Reviews are not intended to replace engineering analysis, and therefore, they cannot be expected to prevent a flight because of a design flaw that management had already determined represented an acceptable risk. . . . *However, a process is only as effective as the responsible individuals make it* [emphasis mine] (150).

Furthermore, their report differed considerably from the presidential commission's in using consistent terminology and in using precise rather than vague or ambiguous language. The result is a set of inferences that are reasonably connected to the evidence and testimony and are not confusing. By contrast, the conclusion section on the contributing causes of the Presidential commission's report states: "The Commission concluded that there was a serious flaw in the decision making process leading up to the launch of flight 51-L [*Challenger*] (104).

We as technical communicators should notice from these contrasting reports stemming from the same body of information that we too can find ourselves facing similar difficult choices. On the one hand, we can choose to communicate using inconsistent or imprecise terminology, which will result in unclear, confusing, or downright misleading impressions on our audience. Or, on the other hand, we can choose to communicate in the opposite manner, using precise terminology consistently, which will result in clearer, more useful messages for our audience. The latter course is the more ethically responsible one but can be the more difficult one because it means we have to lay ourselves on the line. We have to weigh ethical implications, which is always a difficult and problematic activity. We also have to express our judgment clearly, which can often result in disagreement and opposition due to the problematic nature of ethical issues.[2]

---

[2]Roger Boisjoly, the most prominent Morton Thiokol engineer opposing the launch at the L-1 meeting, lost his job because of his outspoken objections and condemnations. After he was demoted and marginalized, he threatened to sue Morton Thiokol under federal whistle-blowing laws. He won his job back, but his working conditions were made so unpleasant that he later left voluntarily.

TWO CRUCIAL SHIFTS IN MEANING

Aside from the governmental reports as documents themselves, the history they trace also teaches us a valuable lesson about technical communication. In two vitally important instances, a shift in meaning occurred having to do with the communication of technical information. One shift occurred slowly, over a period of years, and resulted in a total reversal of meaning of concrete technical information about the charring of O-ring seals. The other shift occurred suddenly on the night before the launch. It involved a total reversal of methodological assumptions from which the information about O-ring charring was perceived. This resulted in the decision to proceed with the launch rather than to postpone it, which the data seemed to warrant.[3]

O-Ring Charring

Before examining these instances, let me first briefly describe the components I will be discussing. The shuttle is composed of three major elements: the shuttle vehicle itself (technically called the orbiter) that is launched into orbit, the huge external fuel tank strapped to the orbiter, and two long solid rocket boosters strapped to the external fuel tank. The solid rocket boosters are like two enormous Fourth-of-July skyrockets, over 100 feet long and 12 feet in diameter. They are so large that they had to be fabricated in cylindrical segments that were then bolted together, one atop the other. These segments (both the shell and the propellant) had to be tightly sealed together to keep the hot, explosive exhaust gases, which should exit from the nozzle, from escaping through the side of the booster. If any leak occurred, it would erode a progressively larger hole for itself like water through a hole in a dike. This would lead to the destruction of the booster and then of the whole vehicle.

Because of the seriousness of possible leaks, the segments were sealed by a set of two rubbery O-rings that were intended to prevent leaks. These were surrounded by a special putty to make the seal more secure and to protect the O-rings from the intensely hot, pressurized exhaust gases. The O-rings were expected never to be exposed to these gases directly, for the gases would quickly eat away the rings and then proceed to destroy the metal segments of the booster shell. These seals were so crucially important that they were given the highest possible "criticality rating" (1-R, critical but redundant) to indicate that they were both critically essential for the safety of any shuttle flight.

From some of the earliest shuttle flights, however, came disturbing data. The spent booster shells are parachuted into the ocean after they have been used up in order to be refueled for another flight. Engineers found that some of the O-rings were charred, though they were expected never to be touched by the

---

[3]These two instances are described in detail in my article "*Challenger* and the Social Contingency of Meaning: Two Lessons for the Technical Communication Classroom."

exhaust gases. There was also clear evidence that the first seal was often not seated properly, effectively making the whole system a single-seal one. This meant that the sealing system was not working as intended and that the booster and the entire flight had been in jeopardy.

This charring was at first rightly understood as a cause for alarm and was reported through the appropriate channels as a danger signal. Because usually only one ring out of a set of two was charred, and then only somewhat, these reports did not stop the shuttle flights. Even if one set were entirely charred through, the other set should help to prevent disaster, it was first thought.

A really strange thing happened next. Over the course of several years, this cause for alarm became reconceptualized, incredibly, as a cause for reassurance of safety! As each shuttle flight went up into space and returned to earth without disaster, even though many of them had some charring of an O-ring, the charring came to be seen in a new light. If a shuttle goes up and comes back without disaster, it was reasoned, everything must be working adequately. Paradoxically, that is, every new instance of charring was taken as confirming that flights can proceed without incident regardless of charring. Eventually, each new instance of charring was thus carefully documented and dutifully reported but not in a way that would cause alarm. And, reciprocally almost, the reports were received without alarm, too.

The congressional report identifies seven instances of "poor technical decisions" concerning the disaster, most of which have to do with the O-rings. It criticizes, for example, decisions made by Lawrence B. Mulloy, manager of the Solid Rocket Booster Project at the Marshall Space Flight Center, who contended that the charring was acceptable because it was within their "experience base." The Congressional report points out the tragic illogic of this reasoning: "In other words, if it broke before and the size of the recent break was no bigger than those before, then there was no problem. Even when the erosion surpassed all previous experience, NASA then went on and expanded its 'experience base'" (50).

Notice that the raw data themselves—the precise, objective information about O-ring charring—remained constant, whereas the representation of what this information meant, how it was to be interpreted, or what should be done in light of it changed totally. Because of this tragic lesson, our technical communication classes should emphasize that we are communicating not just data and facts, but more important, are also communicating what the data mean.

There never was any single, explicit message disseminated to all personnel declaring that henceforth a cause for alarm was to be taken instead as an indication of safety. Instead, over a period of years and through innumerable messages both long and short, a gradual shift in attitude and tone came about, fostered by a number of social, political, and economic forces. (These contextual forces are described in Chapters VI and VII of Vol. 1 of the Presidential commission's report.) They include convenience and simplicity in choosing the sealing system, situational factors such as changes in payload manifest, complications in the training program resulting from mission changes, economic factors compelling quicker turn-around times even though the procedures for turnaround were more involved than anticipated, problematic availability of spare parts, and simple budgetary constraints.

For our purposes here, we need to notice the powerful influence of the entire collection of social forces over a long period of time. This is not to say that this influence was proper or excusable, only that it was real and important. It completely reshaped the attitudes and interpretations of a whole population of bright, earnest people without their even being fully aware of any change. Even the very astronauts slated to fly these missions had read many of the reports of charring yet somehow did not perceive them as drastic danger signals. We as technical communicators should therefore be alert to similar forces possibly at work in our organizational contexts.

We should also notice that these social forces represent values and so have an ethical dimension. Keeping the shuttle on schedule, for example, would not be felt as a pressing need unless it were highly valued. Not only would NASA's future funding possibly be negatively affected, but specific industrial contractual obligations would be left unmet, with serious financial repercussions.

The same holds even for such intangible factors as the image of NASA and its missions in the minds of the American public. NASA's image had sagged since the glory days of the Apollo mission to the moon, which had amounted to a nationalistic conquest and a quasi-military defeat of the Russians by the Americans. NASA wanted to prevent any further erosion of its image and of public support, which would likely result if shuttle launches were postponed for whatever reasons.

Powerful Role of Assumptions

The second instance of a crucial shift in meaning involves a comprehensive technical communication presentation including oral, visual, and written elements. It was an earnest, conscientious, even impassioned presentation urging against launch by several engineers of Morton Thiokol, Inc., the manufacturer of the solid rocket booster, to their management. Their objection was then communicated to NASA management. In the end, tragically, the managers overruled the engineers' strenuous objections and decided to proceed with the launch. The engineer's communications, both prior to this meeting and during it, show how technical communication can be done excellently yet still be nullified by extraneous but more powerful organizational considerations.

The setting is called the L-1 (read "L minus one") Mission Management Team meeting conducted the night before the launch. It brings together the contractors' engineers and managers to decide on a recommendation to launch or to postpone, and NASA management to consider the contractors' recommendations. This is the last major review point from which the launch is approved to proceed or not barring unforeseen developments.

The L-1 meeting for the *Challenger* launch was unusually complicated and protracted because of the issue of O-ring charring as it affected the mission. Several experienced engineers for Morton Thiokol argued against the launch, including Roger Boisjoly and Arnie Thompson. These engineers were well aware of the documentation about the history of O-ring charring and knew how shaky the prevailing thinking about it was. They also knew that the outdoor temperature predicted for the launch was substantially lower than for any previous launch—below freezing for much of the night before the morning launch. They knew, too, that the O-rings become rigid and difficult to seat properly when they are cold. They argued that this colder temperature was "outside their experience base," meaning that whatever the conditions were for other launches did not hold for this one. They concluded that there was a substantial risk of total burn-through or bypassing of the O-rings and hence of the destruction of the shuttle. They recommended against launch.

Ordinarily, the L-1 meeting proceeds methodically but uneventfully. Any significant concerns are raised and then addressed, but usually there are not many concerns and they are not serious enough to postpone the launch. Nearly every aspect of the mission had already been approved several times before at various points in the prelaunch preparation process, so this meeting is more of a double- or triple-check than a completely new checkpoint. The launch preparation is a long, complicated, and expensive process beginning months before scheduled liftoff. Any significant delay would likely develop into a postponement for weeks.

This particular L-1 meeting was unusual in that engineers were arguing strongly against launch, and their Morton Thiokol manager, Robert Lund, did as well. The response of NASA management was unusual, too. The meeting is usually conducted as a telephone conference call linking many different sites across the country; all parties hear what everyone else hears. NASA managers were "appalled" to learn of the engineers' recommendation against launch. "Listeners on the telecon[-ference] were not pleased with the conclusions and recommendations," is Boisjoly's understated testimony (90).

The response to this unexpected and strained exchange was that Morton Thiokol called for an "off-line caucus," meaning that they would discuss the matter among themselves out of hearing of everyone else and off the record. Ten engineers and several high-level managers including several vice-presidents discussed the matter for over half an hour. This is much like the proverbial smoke-filled back room of political campaigns, in which important decisions are made without public scrutiny and, more to the point, without the ethical accountability that comes with public scrutiny. Boisjoly testified about this meeting: "There was never one comment in favor . . . of launching by any engineer or other non-management person in the room before or after the caucus" (92).

The managers insisted to the engineers, however, that this was a "management decision," asserting their prerogative to overrule the engineers' recommendation. And that is just what they did. Lund, who had originally sided with the engineers against launch, was now told by senior management "to take off his engineering hat and put on his management hat" (93). And with this change of perspective, interests, and values came a different interpretation of the data. Rather than being seen as unclear about supporting a recommendation to launch, the data were now seen as unclear about *not* supporting launch; that is, the same technical information was seen as questionable but in the opposite direction.

When NASA management returned to the conference with this new interpretation, they urged that Morton Thiokol management reconsider the data, conclusions, and recommendations yet again from a theoretical basis that supported launching. Morton Thiokol management returned from their reconsideration with a recommendation to launch, which was made official by a telefax under the signature of Joe Kilminster, Vice President of Space Booster Programs at Morton Thiokol.

According to the presidential report, Robert Lund, vice-president of engineering at Morton Thiokol, testified about the changed assumptions:

> We . . . have always been in the position of defending our position to make sure that we were ready to fly. . . . I didn't realize until after that meeting and after several days that we had absolutely changed our position from what we had been before. . . . We had to prove to them that we weren't ready [to fly], and so we got ourselves in the thought process that we were trying to find some way to prove to them that it wouldn't work, and we were unable to do that. We couldn't prove absolutely that that motor [i.e., booster] wouldn't work. . . . It seems like we have always been in the opposite mode (93).

What had happened? Boisjoly clearly describes the crucial change of mind-set that occurred. Remember that there was no change of objective information whatsoever, only a change of perspectives and values. It was a complete reversal of standard assumptions, a total shift in the burden of proof, Boisjoly explained. Because this was not the usual understanding of what this meeting was all about, the engineers were caught off balance and unprepared. More important, even with thorough preparation, they could not have dealt effectively with this radically new argumentative assumption.

Richard Feynman, the renowned physicist and the most distinguished scientist on the presidential commission, was also its most incisive and critical member. His questioning of Boisjoly emphasizes the tragic illogic of the managers' argumentative posture.

Feynman: I take it you were trying to find proof that the seal would fail?

Boisjoly: Yes.

Feynman: And of course, you didn't, you couldn't, because five of them didn't, and if you had proved that they would have all failed, you would have found yourself incorrect because five of them didn't fail.

Boisjoly: That is right (93).

Even more important, the engineers could not have effectively persuaded managers not to launch because the managers had already decided what they were going to do—go ahead with the launch—regardless of what the engineers said. For these reasons, the fateful but erroneous decision to launch was not the result of the engineers arguing and communicating incorrectly or inadequately but of the managers' unwillingness to be persuaded. As Boisjoly summarized this meeting in which he was ignored, "This was a meeting where the determination was to launch, and it was up to us to prove beyond a shadow of a doubt that it was not safe to do so. This is in total reverse to what the position usually is in a preflight conversation or flight readiness review. It is usually exactly opposite to that" (93). Boisjoly and Arnie Thompson continued to argue strenuously against the launch but gave up when nobody listened. The response of management, Boisjoly recalled, was that "nobody said a word" (92).

We should note that not all of Morton Thiokol bought into the new reconceptualization of the data and the new argumentative assumption. A. J. McDonald, director of the Solid Rocket Motor Program at Morton Thiokol, for instance, testified about what he told Morton Thiokol and NASA managers: "I made the statement that if we're wrong and something goes wrong on this flight, I wouldn't want to be the person to stand up in front of [a] board of inquiry and say that I went ahead and told them to go ahead and fly this thing outside what the motor was qualified to. I made that very statement" (95).

The important lesson we should draw from these tragic events is the powerful role in technical communications of things other than the subject matter of technical information. In this case, the technical subject matter and the conscientious manner in which it was communicated were nullified not by new technical information but by the assumptions and expectations of the audience. We like to assume that our audiences are reasonable people open to being persuaded by compelling presentations of convincing information. *Challenger*, however, shows us the realistic limitations of that optimistic assumption.

"SMOKING GUN" MEMORANDUM

In this section we examine the key technical document having to do with the *Challenger* disaster. It is often called "smoking gun" because of its centrality to the events leading up to the disaster. For those unfamiliar with the term "smoking gun," it comes from murder-mystery literature. Imagine the cliché scenario in which a group of people is gathered for a dinner party. Suddenly the lights go out, then a shot rings out. The lights suddenly come on again to reveal that someone has been murdered. The lights also show someone else with a smoking gun in his or her hand. Though no one actually observed the shot itself, the assumption is that the person holding the smoking gun committed the crime.

The prevailing understanding of the O-ring sealing system despite its frequent charring was that it was safe, we saw earlier. Despite this general understanding, several people still refused to buy into this mistaken view. Among the more ethically notable of these people was the Morton Thiokol engineer Roger Boisjoly, mentioned earlier.

One of Boisjoly's technical memoranda (Figure 5.1) is the "smoking gun" of the *Challenger* disaster. The "smoking gun" in this real-life whodunit, however, is a technical document that points the finger of guilt not at its writer but at its readers. (Because this document is entirely in text, I will refer to writing rather than communicating.) Boisjoly's memorandum, we will see, is actually a model of excellent technical writing in many ways, including its ethical dimension. When the writer has done responsibly all that can reasonably be done to write effectively and persuasively, then the burden of responsibility must shift to the readers, the audience of decision makers.

Let us look closely at his memorandum, seeing how it works, what choices went into crafting it, and what the value system behind it consists of. First, read through the memorandum on your own to get a sense of what it is about. The context is reporting about problems with O-ring charring. The document was prepared by Boisjoly as a member of a team established to look into this problem. The team, however, has not been given the time, resources, or opportunity to do its assigned task effectively. One of the most significant aspects of this document is that it combines prominently and effectively a direct, personal voice and emotional appeals with a great deal of highly technical material. It accomplishes all three of its communicative tasks effectively: It is technically informative, rhetorically persuasive, and ethically responsible.

Let us begin at the beginning. The subject line clearly states its subject, as it should, but the writer also takes pains to highlight its significance. The language he uses, "failure" and "criticality," are clear indications of danger, right off the bat, so that no one can mistake the seriousness of the subject. Even a layperson knows that these are serious words and becomes alert and keenly interested.

The first paragraph is brief, as it should be, and clearly states its purpose. Here the writer must engage the reader's interest, and this Boisjoly does. Most technical communication textbooks, keep in mind, emphasize the importance of the first and last statements of a document. The reader is naturally inclined to give these places the most attention. Here the writer expects to engage the reader's interest, and here the writer expresses "the bottom line" significance and calls for concrete action.

To illustrate the significance of the beginning and the ending of a technical document, consider an analogy from everyday social interactions. What you usually remember about a first meeting with another person is the first and the last impression. That does not mean that what happens in between is unimportant, only that these are the natural points of significance for you, just as it is for your reader. Boisjoly, as a good writer, makes the first and last paragraph clear and attention getting. He clearly took his ethical responsibilities as a writer very seriously.

MORTON THIOKOL. INC.

Wasatch Division

Interoffice Memo

31 July 1985
2870:FY86:073

TO: R. K. Lund
              Vice President, Engineering

CC: B. C. Brinton, A. J. McDonald, L. H. Sayer, J. R. Kapp

FROM: R. M. Boisjoly
              Applied Mechanics—Ext. 3525

SUBJECT: SRM O-Ring Erosion/Potential Failure Criticality

This letter is written to insure that management is fully aware of the seriousness of the current O-Ring erosion problem in the SRM joints from an engineering standpoint.

The mistakenly accepted position on the joint problem was to fly without fear of failure and to run a series of design evaluations which would ultimately lead to a solution or at least a significant reduction of the erosion problem. This position is now drastically changed as a result of the SRM 16A nozzle joint erosion which eroded a secondary O-Ring with the primary O-Ring never sealing.

If the same scenario should occur in a field joint (and it could), then it is a jump ball as to the success or failure of the joint because the secondary O-Ring cannot respond to the clevis opening rate and may not be capable of pressurization. The result would be a catastrophe of the highest order—loss of human life.

(continued)

FIGURE 5.1   "Smoking Gun" Memorandum from *Challenger* Disaster, from Roger Boisjoly This memorandum is often characterized as the "smoking gun" of the *Challenger* disaster. Roger Boisjoly strenuously argues that the O-ring charring problem is very serious and requires immediate attention to avoid dire consequences. It is an example of excellent technical communication. It deals with complex technical matters using excellent organization, strong personal voice, compelling emotional language, and a clear sense of ethical responsibility.

From *Report of the Presidential Commission on the Space Shuttle* Challenger *Accident*, Vol. I, Appendix D., p. 249.

An unofficial team (a memo defining the team and its purpose was never published) with leader was formed on 19 July 1985 and was tasked with solving the problem for both the short and long term. This unofficial team is essentially nonexistent at this time. In my opinion, the team must be officially given the responsibility and the authority to execute the work that needs to be done on a non-interference basis (full time assignment until completed).

It is my honest and very real fear that if we do not take immediate action to dedicate a team to solve the problem with the field joint having the number one priority, then we stand in jeopardy of losing a flight along with all the launch pad facilities.

R. M. Boisjoly

Concurred by:
J. R. Kapp, Manager
Applied Mechanics

COMPANY PRIVATE

FIGURE 5.1   Continued

Boisjoly evokes the attention of his managerial audience by his tone and posture. He states that he is addressing a managerial audience, which his reader will immediately grasp as himself or herself specifically rather than just a disinterested general audience. Boisjoly also states his intention that this memorandum will make that audience "fully aware" of a problem. The reader reads this as saying in effect, "You have been warned." Even a layperson recognizes the potential "I told you so" should the writer's message go unheeded and becomes alert to avoid that possibility.

To illustrate Boisjoly's careful and purposeful use of language, notice that he deliberately decided to include certain words that he could have left out. The result would have been a "technically" true and correct communication but without the compelling persuasive force of the actual document. Note in the first paragraph that he could have left out "fully," "seriousness of the," and "problem." This would leave a relatively limp but technically correct sentence that does not even acknowledge that there is a *problem*, much less a *serious* one. Boisjoly instead chose to include these emotionally powerful words.

The second paragraph is largely a description, which is fairly common in a technical report about a problem, before moving on to a solution. Boisjoly describes the past and current state of affairs. He could have left out several key words here, too: "mistakenly," "drastically," and "never." Without the "mistakenly," we would have a true description of the current state of affairs that would sound fairly positive. With the "mistakenly" added at the very beginning of the sentence, however, the reader knows right off the bat that something is amiss. The "drastically" adds emotional urgency to what otherwise would be a flat statement of fact. Notice too that "never" has a sense of grim finality that would be lost had Boisjoly said flatly, "the primary O-ring did not seal."

The third paragraph is a hypothetical but plausible outline of how the scenario described in the second paragraph could become a very real problem, again a fairly typical movement in a technical report.[4] It contains a good deal of highly technical information, which is presented clearly, objectively, and unemotionally. Boisjoly chooses, however, not to leave it at that. Notice that this paragraph contains an interesting mix of technical terminology with an everyday expression. Boisjoly deliberately chose to include "jump ball" because of its rhetorical effect. This colloquial term would be familiar to his audience and understood at a visceral and emotional level. Anyone who has played basketball knows that a vitally important matter, possession of the ball, in a jump ball is beyond the control of the participants: It could go either way. He could have said flatly, "the probability of effective sealing would be 0.50," which would be technically correct but would fail to grab the reader's attention.

Notice, more importantly, that he could have left off the entire last sentence and still had a technically correct paragraph but, again, without the powerful emotional impact and ethical thrust. Within this last sentence, furthermore, he uses very strong, completely unambiguous language: "catastrophe," "highest," and "loss of human life." Notice too that the phrase "loss of human life" is a restatement of what "the highest order" means, just in case the reader does not understand this immediately. This restatement, some might say, is an unnecessary overdetermination or redundancy. In this case, though, because of the great seriousness of the topic, this overdetermination is the wise and ethical choice.

This sentence also takes the paragraph beyond the simple description of facts to the drawing of an explicit conclusion that spells out the full meaning and significance of the technical facts just presented. This way the reader does not have to infer anything; the technical communicator takes the responsibility of spelling it out explicitly for the reader. It is generally a good idea to go this "extra mile" in your technical communications, drawing an explicit conclusion so that the reader is absolutely clear about what your meaning is.

The fourth paragraph outlines the organizational situation and complains that the writer feels that his organizational and ethical responsibilities cannot be fulfilled under the present situation. Boisjoly could have left out "essentially

---

[4]As a technical note, there were two kinds of O-rings on the shuttle boosters. The nozzle, which could be maneuvered slightly, had "nozzle O-rings." The segments of the body of the booster had "field joint O-rings."

non-existent" and still had a valid informative paragraph but a weak one. To emphasize ethical responsibility as a primary concern of the writer (and implicitly that it should also be so for the reader), he includes the phrase "must be officially given the responsibility and authority." Otherwise, the sentence would have read something like this: "The work of this unofficial team needs to be done on a non-interference basis." Notice too that, as in the preceding paragraph, the writer takes pains to restate his point, restating what "non-interference basis" means. Restatement makes sure that the reader understands your meaning.

The final paragraph is the clincher. Notice that this entire paragraph could have been left off. The resulting document would have been weak but not grossly incorrect as a technical document. Some writers might prefer not to put themselves on the line. They might prefer not to call strongly for concrete action, hoping not to rock the boat. Boisjoly, however, does not leave it at that. If the reader still has not gotten his message that something terribly serious is going on, the writer makes one final attempt to elicit the reader's active concern. Notice that he does this in a highly personal and emotional voice in active language. He could instead have left off, "It is my honest and very real fear" and still have been correct. Instead, he chose the emotion of "fear" and the moral and ethical value of "honest(y)." Furthermore, he takes his responsibility personally and expresses himself so with "my." This highly personal and emotional voice, were it to occur in any sort of writing, is guaranteed to be noticed, especially so in technical communication.

Boisjoly could also have left off the entire last clause, "then we stand in jeopardy of losing a flight along with all the launch pad facilities," replacing it with something mild like, "then we might possibly encounter a serious mishap." Instead, again, he takes the extra step of spelling it out explicitly and in detail for the reader so that his meaning is absolutely clear.

Most technical communicators would agree that this is an excellent piece of writing. It deals with highly technical material but does so using personal voice, active constructions, and everyday language where it would be most effective. It also uses emotional and ethical appeals, not to mislead or seduce the audience but to gain its genuinely needed concern.

The technical communicator has done the job of communicating well. If the audience does not get the message or chooses not to act on it, then that must be the responsibility of the audience but not the technical communicator. And that is all you can ever do, doing your best to hold up your ethical end of the communicative transaction. Communication is a two-way street, and the audience has its own, separate responsibility to hold up its end of the transaction.

GRAPHICAL IMAGES

We have seen in the previous sections that apparently objective, concrete technical information such as "flightworthiness" and "safety" are actually shifting concepts that can change depending on personal and social context. Several

SRM-16A Primary Nozzle O-Ring—Right Notch (Chart A-12)
54 Deg Location Looking Towards 90 Deg

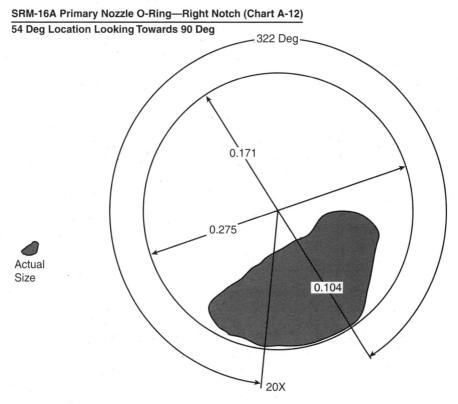

FIGURE 5.2   Slide Showing O-Ring Charring This slide from the flight readiness review process shows visually and numerically the extent of charring in a cross section of the nozzle O-ring of shuttle flight 16A.

From *Report of the Presidential Commission on the Space Shuttle* Challenger *Accident*, Vol. II, Appendix H., Chart 106, page H-54.

graphical images from the *Challenger* investigations make a similar important point about technical information: It is more than just data or numbers.

The two sample slides we will examine reveal two points. First, data in the form of numbers can be almost irrelevant for technical information purposes because they can, depending on circumstances, be beside the point. Instead, what the numbers *mean* is the point. Second, numbers do not always determine their own meaning, and so collecting more and better numbers would not necessarily alter or improve meaning.

Figures 5.2 and 5.3 show several actual overhead transparency slides that were used in the flight readiness review process for various shuttle missions just before the *Challenger* disaster. They show both visually and in numbers the extent of charring at several points on certain O-rings from various missions. Keep in mind that the O-rings themselves are rubberlike rings of about the

Past History Comparison (Chart B-3)

Nozzle O-Ring Erosion Patterns (Optical Comparator)

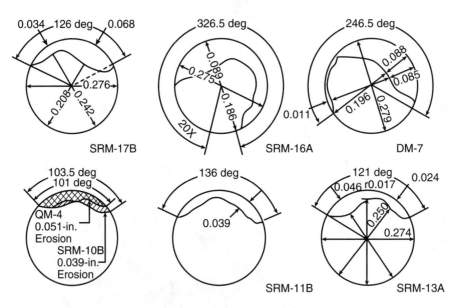

FIGURE 5.3   Slide Showing History of O-Ring Charring on Several Flights This slide from the flight readiness review process shows visually and numerically the extent of charring in cross sections of O-rings from several shuttle flights before the *Challenger* disaster.

From *Report of the Presidential Commission on the Space Shuttle* Challenger *Accident*, Vol. II, Appendix H, Chart 113, page H-58.

same thickness as a lead pencil. Though the diameter of the cross section of the ring is about 0.275 inches, the total diameter of the entire ring is about twelve feet. The crucial fact is that the O-rings were never supposed to be directly exposed to the rocket gasses at any time: *There should never be any charring whatsoever.*

The slides show magnified cross sections (note "actual size" and "20X" magnification and the "0.275" [inches] cross-sectional diameter in Fig. 5.2) of the O-rings at various points. Figure 5.2 dramatically shows in visual terms that most of the O-ring had been eroded by the hot gasses inside the solid rocket booster. Clearly most of the O-ring is gone. Figure 5.2 also shows in great numerical detail the exact depth, location, and expanse of the charring and the depth of the remaining O-ring material. This charring occurred at the 54-degree location on the circumference of the ring as it was positioned on this particular booster.

Figure 5.3 shows part of the history of the charring of O-rings for several past shuttle flights such as "SRM-17B." The slide shows that they were all seriously eroded and shows numerically the exact extent of the charring. (As visual aids, both slides are less than ideal. They do not include scale or units and several times place numbers with different implied units next to one another, for example, and erosion is represented in most drawings as blank space but in one drawing as cross hatching.)

The point about these slides is that one does not have to be a rocket scientist to see that the O-rings are seriously eroded. Given the knowledge that the O-rings should *never* be eroded at any time to any degree, nearly anyone can see that something is seriously wrong here. Nevertheless, as explained in earlier sections of this chapter, somehow this information was reconceptualized or reinterpreted in such a way as to suggest safety!

The crucial conclusion we should draw from these images is that the numerical data and even the images themselves visually are not the point, really. Would it have made any difference if the measurements of the charring were carried to 5 significant digits rather than just 3 (0.171xx vs. 0.171)? Would it have made any difference if the entire history of all O-ring charring instances were presented in a series of slides rather than just six on one slide? Or if all the radial angles were measured with greater precision (101.xx degrees vs. 101 degrees)? The answer obviously is, No. The real point is what the images and the numbers *mean* and how they are to be interpreted. Given different assumptions or interpretative schemes, the numerical data apparently can mean different things or be taken in different ways. And collecting more and more data and trying to be more precise with the numbers would practically be beside the point because they would not make any difference. Instead, how the numbers are understood—what conceptual framework is brought to our viewing of them—is really what was most important as far as the lives of the *Challenger* crew is concerned. Therefore, the data or numbers do not determine their own meaning. Instead, people determine what numbers mean.

ETHICAL APPRAISAL

Let us see how these technical communications would be appraised from the various ethical perspectives we have discussed.

Aristotle.   As for Roger Boisjoly's memorandum and his strenuous arguing during the L-1 meeting, Aristotle's perspective would clearly commend him as completely ethical in insisting on the good, true, and right. Certainly his technical information was true, he was arguing for the right interpretation of that information, and he sought to do good for the profession, the agency, the mission, and the astronauts. For Aristotle, behaving ethically is a matter of cultivated virtue and personal character. Ethics is therefore an expression of character, and so ethical behavior often occurs in a context in which other virtuous facets of one's character are also expressed. Such was the case with Boisjoly, who showed his honesty in insisting on the truth, his courage in opposing the view of those more powerful than himself, his honor in defending the accepted principles of his profession, and his selflessness in trying to protect the lives of those who depended on him. This case illustrates very clearly how ethics is not a one-dimensional concept for Aristotle.

Regarding the two governmental reports, the Aristotelian appraisal would be mixed in all likelihood. The ambiguous, confusing language of the presidential commission's report and its unwillingness to focus on personal responsibility would not warrant being called ethical. It even borders on obfuscation (deliberately making things unclear), which seeks to avoid the true, good, and right, and so seems clearly to be unethical. The congressional committee's report more definitely sought to clarify rather than to confuse, particularly regarding personal responsibility. This concern for revealing the truth and identifying responsibility would have to be considered highly ethical.

The interplay between these two reports is itself interesting from an Aristotelian perspective. Recall that Aristotle advocated vigorous, open debate in order to resolve problematic issues because he thought that this adversarial interaction would allow the natural goodness and rightness of one side to assert itself over the other and to prevail. That seems to be what is happening here right now as we debate within ourselves which of the two different (and almost opposing) governmental reports we will believe to be truer, righter, and better than the other.

Kant. A sense of duty, of doing the right thing regardless of personal consequences, is the basis of the Kantian perspective. In the case of Roger Boisjoly's memorandum and his presentation at the L-1 meeting, he obviously accepted and enacted his ethical responsibilities even though he was communicating a message that his audience did not want to read or hear. He suffered for doing his duty, losing his job and his colleagues, even though his ethical judgment was vindicated after the fact. And he will always be remembered in the fields of technical communication and engineering for his ethical responsibility. Boisjoly continues to be invited to lecture on engineering ethics.

As for the two contrasting governmental reports, the congressional committee clearly accepted their responsibility for doing what they thought ethically best, and that was to focus on judgment and personal ethical responsibility. The presidential commission, on the other hand, seems to have deliberately avoided focusing on personal ethical responsibility, even though the evidence and testimony suggest the need for that focus. As a result, the technical report of the presidential commission shows confused language, flawed reasoning, misleading conclusions and recommendations, and so seems not to have behaved ethically toward its audience.

Utilitarianism. The calculated balancing of good against harm underlies the utilitarian perspective. In the case of Boisjoly's memorandum, what constitutes the good effects and what constitutes the harmful effects is rather unclear. No one doubts that saving the lives of the astronauts is a good idea, but, it might be argued, they knew they were involved in a very risky business and willingly accepted that risk. That risk involves the near-certainty that some astronauts will die at some time. That risk is, however, outweighed by the greater good to the greater society of advancing the technology, exploring the unknown, adding to scientific knowledge, and maintaining national security. Not only is this the same calculation that NASA and the federal government made collectively, but so does each of the astronauts individually. This calculation is even more legitimate in a historical sense if we recall that the primary motivation for space missions among the superpowers was national military security. In wartime, even during a "cold" war, we expect that some people will pay a price that would be unacceptable during peacetime. The ethical rub, however, is that some astronauts were being exposed to greater risk than they were aware of and to risks that could have been minimized but were not.

In the case of the contrasting governmental reports, one might argue that, when the presidential commission avoided addressing ethical personal responsibility, it was doing so out of a desire to serve the greatest good for the greatest number. According to this line of reasoning, it represented the disaster as only a regrettable accident in order to maintain the reputation of NASA and to ensure generous funding for its missions and for the sake of national security. The difficulty with this line of reasoning is that presumably the legislative branch, the U.S. Congress, is just as concerned about national security as the executive branch, presided over by the President. Yet the congressional committee did not feel that national security concerns were compromised by focusing on personal ethical responsibility.

Feminist and Ethics of Care.   The political dimension is important in feminist criticism. In overriding the engineer's well-informed recommendation against the launch, management behaved in an authoritarian way, insisting on reserving the final judgment for themselves, insisting on the greater power of themselves over the engineers, and insisting that they could even turn the tables by inventing a new argumentative assumption for the L-1 meeting.

Caring concern for others would require that those involved in communicating and making decisions about the *Challenger* mission do so with the safety of the crew paramount in their minds. This seems not to have been the case as the safety of the crew was overridden by concerns about such things as the flight schedule, NASA's image, the magnitude of funding for NASA in the future, the perceived thoroughness of engineering knowledge, management's authority, and personal reputation.

Therefore the decision to launch seems clearly unethical from the perspectives of feminist ethics and an ethics of care.

CONCLUSION

The *Challenger* disaster example shows how values and ethical judgment play a key role in the communication of even highly technical information. This holds true also for the two governmental reports, which supposedly were aimed at getting to the root of what "really" happened. Even then, the values, interests, purposes, and assumptions of the communicators played a vitally important role in shaping their reports and in shaping how technical information was gained, represented, and interpreted. Within the history of events leading up to the disaster, there were at least two important ways in which nontechnical factors (such as values and methodological assumptions driven by values) shaped the way technical information was represented and interpreted. These forces even completely reversed the way the information was understood. The same holds true for the "smoking gun" memorandum, which, though exemplary in its persuasive appeals and execution of ethical responsibilities, was undone by the values and interests of a different, more powerful organizational group.

Topics for Papers and Discussion

1. The *Challenger* disaster in 1986 and the near-meltdown of the nuclear reactor at Three Mile Island in 1979 are the two most widely cited technological disasters of recent times in the United States. They are widely cited because they were so dramatic and so serious (though little radiation escaped at Three Mile Island (TMI), there was a very real potential for a great deal more).[5] They are also cited because of the ethical dimensions of the technical communications associated with the disasters. In the case of the *Challenger* "smoking gun" memorandum, we saw how Roger Boisjoly carefully enacted his ethical responsibilities in his technical communications, though his readers seem to have been considerably less responsible.

The Three Mile Island case is quite different from *Challenger*. Here the writer seems to have avoided enacting responsibility effectively. This left the reader without a clear, compelling document on a vitally important topic. The document usually identified as the "smoking gun" in this case comes from within the Babcock and Wilcox corporation, which manufactured and provided many of the operating instructions for the reactor and its associated steam equipment. It was written about seven months before the disaster and refers to lessons learned from an accident similar to TMI that occurred just two years before. Though this document is highly technical, dealing with nuclear engi-

---

[5]Serous though Three Mile Island was, especially potentially, it did not release substantial amounts of radiation into the environment. The gigantic nuclear disaster and actual meltdown at Chernobyl in the former Soviet Union released about one million times as much radiation into the environment as Three Mile Island.

neering and power generation, one does not have to fully understand the terminology to follow its drift or to understand the technical communication decisions that went into drafting this document. These writing decisions in turn suggest certain values operating behind this document, whether consciously or unconsciously. Locate this document and perform an ethical analysis of its features, along the same lines as for the *Challenger* memorandum you just read. Pay particular attention to word choice, the focus of each sentence, the placement of key information, and any cues to danger or urgency.

The document, a memorandum from D. F. Hallman to B. A. Karrasch dated August 3, 1987, can be found as Appendix N of the *Report of the Office of Chief Counsel on the Role of the Managing Utility and Its Suppliers* (see Figure 5.4 at the end of this chapter). Note that this document is *not* the presidential commission's report on the disaster itself, which is largely a description of actions during the disaster, but sort of an addendum to it. As the title page explains, "This document is solely the work of the Commission staff and does not necessarily represent the views of the President's Commission or of any member of the Commission." This related but not representative report searches for explanations, causes, and responsibilities. The actual report to the president, on the other hand, seems to avoid discussing responsibility and operates at the level of narrative and description only.

Clarity of technical communication is the focus of many of their findings. About the fumbling actions in the control room at the time of the near-meltdown, for instance, it explains: "To attribute the error solely to the operator is an oversimplification. More importantly, it begs the underlying issues of who supplied and shaped the analytical, procedural, and intellectual tools relied on by the operators and their supervisors in the critical first hours of the accident" (90). Pages 108–9, 127, 195, and 196 provide other strong statements, among many in the report. Regarding the impact on the reader of the Hallman memorandum, see pages 134–36. Read the section on "areas deserving further investigation" beyond the scope of this particular report (pp. 198–99). How many of these items are related directly or indirectly to technical communication, and in what ways?

If you wish to further explore the Three Mile Island near-meltdown, look into the deliberate falsification of reports about leaking reactor coolant at Unit 2, the one that suffered a near-meltdown. These reports covered the six-month period up to the day of the accident as the leakage rate grew progressively higher. Though the U.S. Nuclear Regulatory Commission found no evidence of criminal misconduct, it did find that there was a pattern of the systematic, deliberate communication of false and misleading technical information. In response to federal charges, Metropolitan Edison, the owners of the Three Mile Island plant, pleaded guilty to one count and no contest to six other charges.

2. The technical documents generated by the investigation of the *Challenger* disaster provides several striking illustrations of ethical responsibility in highly technical communications. One of the most powerful and clear of these comes from Richard Feynman. Feynman, who died in 1988, was one of America's foremost physicists and a beloved teacher of science. He was also the most distinguished scientist on the presidential ("Rogers") commission investigating the disaster. The presidential commission wrote a voluminous report, parts of which are examined in this chapter. As we saw, some of the conclusions and recommendations seem not to follow from the evidence and testimony and seem to be inappropriately mild in addressing ethical concerns (especially in contrast to the congressional committee's report).

Feynman felt dissatisfied with the overall thrust of the commission's final published report. He was so ethically troubled, in fact, that he wrote his own separate statement about the disaster and its investigation of it, which directly challenged or qualified some of the statements in the commission's report. Feynmann's independent statement, motivated by his ethical concern, was intended to represent truthfully the underlying problems that led up to the disaster. It also indicates serious problems with the meaning and usage of key technical terms. These terms appear in usage to be definite in their meaning but are actually used in the commission's report either loosely or in highly misleading ways. Feynman's technical document is titled "Personal Observations on Reliability of Shuttle" and was published as Appendix F in Vol. II of the presidential commission's report. It deals with highly technical information yet in a manner that explicitly highlights ethical concerns.

Locate and study this technical document. Identify key passages that communicate the ethical concerns in this technical communication, which is an excellent illustration of highly technical information dealt with using strong ethical responsibility and an accessible style. Characterize how effectively the document interweaves technical and ethical issues. Identify some of Feynman's stylistic choices, and characterize their suitability and effectiveness. (Feynman was famous for his lucid style and for making difficult subjects readily accessible. His textbook on quantum electrodynamics is still one of the standards on this subject, and videotapes of his famous lectures are still popular even decades after they were first delivered.) Does his style support his ethical message?

3. How much responsibility does a technical communicator have for how a technical document ultimately is read and understood by the audience? One of the tragically paradoxical findings of the *Challenger* investigations was that many of the shuttle astronauts received, read, and later acknowledged having read key reports about problems with the O-rings, yet had not been alarmed. They even acknowledged that though formal documents that they had read had stated that the criticality rating for O-rings was changed from 1-R to 1, they nonetheless still continued to think of the O-rings as a 1-R device. (1-R means vitally important but redundant, that is, a dual system with a built-in backup; just 1 means vitally important but with no backup.) After the criticality change was officially announced, later documents read by the astronauts referred to the O-rings as having a 1 rating. It seems that the message was accurately and properly communicated, but the audience refused to acknowledge the full significance of the communication.

This raises the question, does ethical responsibility end after the original message is communicated? Does the ethical burden then lie solely with the receiver of the communication? Are you responsible if your audience chooses to ignore or disregard your technical communications? In the case of Roger Boisjoly, after his initial communications were ignored or inadequately responded to, he continued to send the message repeatedly and in even more attention-getting language. What do you think are the limits to which you should go to ensure that your audience fully understands and accepts the meaning you are communicating? And why do you feel these limits are warranted? Write a brief report on these questions.

4. Many technical communication articles have been published about *Challenger*. They point to many different explanations about why the disaster happened and the role of communication in it. One of the more frequent explanations, building on the report by the presidential commission about contributing causes such as economic and political pressures, emphasizes the social or organizational forces at work in the overall communication situation. This sort of explanation is called "social constructionist" because it emphasizes how knowledge is "constructed" by social forces rather than being just simply discovered.

Though there is a great deal of validity attributed to social constructionist explanations throughout the literature of technical communication, explanations of this sort can be more complicated and problematic than they might seem. Read Dorothy Winsor's widely cited article, "The Construction of Knowledge in Organizations: Asking the Right Questions about the *Challenger*." Summarize its argument and conclusions, and discuss its general significance for technical communication studies. Do the same for Charles Harris, Jr.'s article "Explaining Disasters: The Case for Preventive Ethics."

5. Another widely cited article on the role of social and organizational forces in the *Challenger* disaster is Patrick Moore's "When Politeness is Fatal: Technical Communication and the *Challenger* Accident." This social constructionist perspective on *Challenger* emphasizes the very real dilemma that technical communicators can face in communicating information, conclusions, and recommendation that differ from what is expected or desired by management. Politeness and the sensitive accommodation of the expectations of an audience are generally valuable communication techniques. The efforts of a technical communicator to make a message acceptable (to both sender and receiver) can, however, also lead to misinterpretations by the audience, with grave consequences. Read Moore's article. Summarize its argument and conclusions, and discuss their general relevance for other technical communication situations such as you might encounter.

```
BABCOCK & WILCOX COMPANY
POWER GENERATION GROUP

TO:      B.A. Karrasch, Manager, Plant Integration        cc: E.R. Kane
                                                              J.D. Phinney
FROM: D.F. Hallman, Manager,                                  B.W. Street
         Plant Performance Services Section (2149)            B.M. Dunn
                                                              J.F. Walters
CUST:                                                  File No.
                                                       or Ref.

SUBJ:    Operator Interruption of High         Date: August 3, 1978
         Pressure Injection (HPI)

References: (1) B. M. Dunn to J. Taylor, same subject, February 9, 1978
            (2) B. M. Dunn to J. Taylor, same subject, February 16, 1978

References 1 and 2 (attached) recommend a change in B&W's philosophy for
HPI system use during low-pressure transients. Basically, they recommend
leaving the HPI pumps on, once HPI has been initiated, until it can be
determined that the hot leg temperature is more than 50°F below Tsat for the
RCS pressure.

Nuclear Service believes this mode can cause the RCS (including the pressurizer)
to go solid. The pressurizer reliefs will lift, with a water surge through the
discharge piping into the quench tank.

We believe the following incidents should be evaluated:

1. If the pressurizer goes solid with one or more HPI pumps continuing to
   operate, would there be a pressure spike before the reliefs open which could
   cause damage to the RCS?

2. What damage would the water surge through the relief valve discharge piping
   and quench tank cause?

To date, Nuclear Service has not notified our operating plants to change HPI
policy consistent with References 1 and 2 because of our above-stated questions.
Yet, the references suggest the possibility of uncovering the core if present HPI
policy is continued.

We request that Integration resolve the issue of how the HPI system should be
used. We are available to help as needed.

                        /s/ _____
                                D. F. Hallman
DFH/fch
```

FIGURE 5.4 **"Smoking Gun" Memorandum from Three Mile Island** This
memorandum from Babcock and Wilcox Company, which manufactured and helped
operate the nuclear reactor of Three Mile Island, is often considered the "smoking
gun" in this disaster. It discusses possible problems with certain procedures in certain
unusual conditions. "High Pressure Injection" is meant to cool the hot reactor quickly
in an emergency. The "RCS" is the reactor cooling system. The memorandum is an
example of poor technical communication and poor ethical awareness.

From *Report of the Office of Chief Counsel on the Role of the Managing Utility and Its Supplies* (also known
as *The Staff Report to the President's Commission on the Accident at Three Mile Island*), Appendix N, p. 227.

REFERENCES

Dombrowski, Paul M. "The Lessons of the *Challenger* Disaster." *IEEE Transactions on Professional Communication* 34/4 (1991): 211–61.

_____. "*Challenger* and the Social Contingency of Meaning: Two Lessons for the Technical Communication Classroom." *Technical Communication Quarterly* 1/3 (1992): 73–86.

_____. "Can Ethics Be Technologized? Lessons from *Challenger,* Philosophy, and Rhetoric." *IEEE Transactions on Professional Communication* 38/3 (Sept.1995): 146–50.

Feynman, Richard P. "Personal Observations of Reliability of Shuttle." *Report of the Presidential Commission on the Space Shuttle Challenger Accident.* Vol. II. Appendix F. Washington, D.C.: GPO, June 6, 1986.

Harris, Charles H., Jr. "Explaining Disasters: The Case for Preventive Ethics." *IEEE Technology and Society Magazine* 13 (Summer 1995): 22–27.

Herndel, Carl G., B. A. Fennell, and Carolyn R. Miller. "Understanding Failures in Organizations: The Accidents at Three Mile Island and the Shuttle *Challenger.*" *Textual Dynamics of the Professions: Historical and Contemporary Studies in Writing in Professional Communities.* Madison, Wisconsin: University of Wisconsin Press, 1991.

Moore, Patrick. "When Politeness Is Fatal: Technical Communications and the *Challenger* Accident." *Journal of Business and Technical Communication* 6/3 (July 1992): 269–92.

Pace, R. C. "Technical Communication, Group Differentiation, and the Decision to Launch the Space Shuttle *Challenger.*" *Journal of Technical Writing and Communication* 18/3 (1988): 207–20.

Report of the Office of Chief Counsel on the Role of the Managing Utility and Its Suppliers. (Also known as the Staff Report to the President's Commission on the Accident at Three Mile Island.) Stanley M. Gorinson, Chief Counsel, and Kevin P. Kane, Deputy Counsel. Washington, D.C.: GPO, October 1979.

United States. Presidential Commission on the Space Shuttle *Challenger* Accident. *Report to the President by the Presidential Commission on the Space Shuttle Challenger Accident.* 86-16083. Washington, D.C.: GPO, 1986.

_____. Congress, House. *Investigations of the Challenger Accident: Report of the Committee on Science and Technology.* House of Representatives, Ninety-ninth Congress. 87-4033. Washington, D.C.: GPO, 1987.

Winsor, Dorothy. "The Construction of Knowledge in Organizations: Asking the Right Questions about the *Challenger.*" *Journal of Business and Technical Communication* 4/2 (1990): 7–20.

■ ■ ■ ■ ■

CHAPTER 6

TOBACCO AND DEATH
WHEN IS A CAUSE NOT A CAUSE?

If the avoidable sudden and violent death of seven astronauts can be called a tragedy, what word can describe the avoidable, slow, and misery-filled death of millions? This is not only a rhetorical question but a very real one when considering the ethical dimensions of technical documents that relate to the manufacture and advertising of cigarettes in the United States since the 1950s. By all reputable medical, scientific, and governmental accounts, literally millions of people have died from having smoked cigarettes.[1] These deaths resulted from specific diseases caused by smoking or from the aggravation or complication of other disorders or simply from a shortened life span. The tragedy lies in the entirely preventable nature of these deaths and in the strained resistance of the tobacco industry to acknowledging its ethical responsibilities.

This tragic dimension, in fact, is the reason this chapter more than some others seems to take a particular side on the "debate." In reality, there was no debate. That is precisely the reason so many governmental authorities throughout the country are suing the tobacco industry. The industry knew full well, these authorities contend, that they were causing serious illness and death on a tremendous scale. Also more so than in other chapters, this one uses "sophistic" only in its traditional sense of contrived contention where none is warranted, of "making the worse case seem the better," of the single-minded pursuit of winning rather than pursuit of what is true and good, and of denying obvious realities. There are, of course, other valid uses of the term "sophistic." In recent years the sophists have been recognized as performing valuable, innovative intellectual activities in showing the power of language to shape ideas and opinions and even knowledge itself. They also demonstrated the effective power of language to challenge prevailing structures of power and authority.

---

[1]The World Health Organization estimates that in the developed countries alone, about two million people died from smoking just in 1995, and a total of about sixty million in the years 1950–2000. For the developing countries, the figures are still being accumulated but are known to be steadily climbing.

In the case of the *Challenger* disaster, part of the reason we are so outraged by the ethical lapses in technical communications associated with it is that the seven deaths were sudden, violent, and witnessed by millions (and witnessed by still millions more later on video playback). Clearly something happened that should not have happened. In the case of death from smoking, however, the appearances are very different. When someone contracts lung cancer, death follows months or years of painful, progressive debilitation and destruction. These deaths are often lonely, witnessed only by nurses and perhaps a few family members. There are no video records broadcast to the world of the death itself or of the years of misery preceding it. There is no public drama here, just the very real personal tragedy that is unbearable for others to watch. We all have seen attractive billboard advertisements of the vigorous, handsome Marlboro Man, for instance, but no one except close family watched at his deathbed as his body was consumed by cancer from decades of heavy smoking.

In addition to deaths from specific diseases, it is clear that steady smokers will in general lose from ten to twenty years from their nonsmoking life expectancy, regardless of the specific disorder from which they die. Yet how can one visualize in concrete, dramatic terms a premature death from an unspecified disorder? It is a far cry from the dramatic visual impact of the *Challenger* exploding before our eyes but is no less real and no less important.

In this chapter we will see how technical and scientific information about smoking and disease was deliberately controlled, manipulated, and misrepresented to various audiences by the tobacco industry. These audiences include the public, the government, the scientific community, and even the tobacco industry itself. We will see, too, how even the strict meaning of a single technical word can become the focal point of a complex web of misleading arguments spanning several decades.

As in other chapters, technical information will not be distinguished from scientific information because the distinction is not necessary or useful for our purposes. Both technical and scientific information are specialized, usually quantitative knowledge, and both were used in similar ways for similar reasons in the interest of corporate and personal gain. This was done in many ways such as disguising self-serving posturings as technical facts, by disguising grim realities as pleasurable indulgences, and by disguising knowledge as beliefs. In all these activities, the representation of technical information and the active shaping of its meaning together with the validity of inferences drawn from it played key roles. In this case, unfortunately, these activities had the effect of misleading the public and evading ethical responsibilities. Because these activities are also intrinsic to most ordinary technical communication situations in other industries, too, we need to be alert to the values and motives driving our technical discourse and be wary of slipping into ethical lapses.

First we ethically examine a quibble about causation, then specific examples of documents showing the misrepresentation of technical information, then another quibble about a single word, and finally several graphical images.

CAUSE

There are two other major reasons why the death of millions from lung cancer due to smoking does not outrage us as much as the death of the seven *Challenger* astronauts. Both reasons relate directly to this chapter and have to do with the cause of these deaths. The first reason concerns probabilities and populations rather than certainties and individual persons. In the minds of most people, causation is a fairly direct and mechanical matter. Heating a snowflake causes it to melt, always, everywhere, and for everyone. It cannot be otherwise. The shattered *Challenger* vehicle fell to earth because of gravity and could not have done otherwise. This is a narrow sense of the concept of causality, but it is the one we all know intuitively.

There is, however, an equally legitimate meaning of the word *cause*, a statistical one. Statistical causation means a probability of what will happen in a population or group but says nothing about particular individual cases—populations are precisely what statistics and probability are all about. Every one of us is interested first and foremost about what will happen to ourselves as individuals, but statistical inference is not about particular individuals. Other people, including groups of others known as "populations," on the other hand, are of only secondary interest to each of us personally. Unfortunately, we all too often fail to realize that each of us is a member of a "population" of other people to someone else. We are already among someone else's "other."

With statistical causes, the connection between cause and effect is often not immediately seen and not intuitively obvious. It might take years or decades for a cause to work its ultimate effect on a body, as it does in the case of smoking and cancer. But though the time frame is much longer for statistical understandings of causation, the causal connection is not less real. Those old enough to remember leaded gasoline will recall that the connection between tetraethyl lead added to gasoline and lead poisoning took a good deal of time to come to light. The causal connection was demonstrated only after many years of correlating the occurrence of the signs of poisoning with proximity to gasoline combustion products. Because of this statistical connection, nearly all our gasoline these days is unleaded. For similar statistical reasons, the federal EPA in 1999 sought to ban the fuel additive MTBE. The same holds for the ecological effects of DDT and other pesticides. Though the connection between cause and effect is a long thread, it is a very real one.

In the case of *Challenger*, on the other hand, the connection between cause and effect was close in time. The immediate, material cause was the burn-through of the O-ring seals, which brought about its ultimate effect—destruction of the vehicle—in less than two minutes. We even have vivid pictures of the black puff of smoke at the exact moment of liftoff of the O-ring seal as it failed, leading to the fiery explosion just two minutes later.

The second reason also has to do with causation but from another angle and an unethical one, as we will see. From the very beginning of serious public health reports about the real medical consequences of smoking and other forms of tobacco use in the 1950s, the tobacco industry has engaged in an aggressive program of misinformation, obfuscation, denial, and opposition that has clouded the connection between smoking and disease in the minds of the public. The tobacco industry, for example, has continually asserted the existence of a "controversy" about whether smoking really "causes" disease, when in fact there was no controversy except of their own manufacture.

The language just now used—misinformation, obfuscation, denial, opposition—is clear and strong. It is not qualified by adding, for example, "the possibility exists that obfuscation occurred" or "the denial of causation appears to have been not well founded." The direct language is appropriate because that is the same sort of language that various governmental entities in their law suits have used and are now using about the tobacco industry and winning either favorable judgments or massive concessions and settlements. It is also the language that medical and public health officials in the United States and throughout the world have used about the tobacco industry.

Statistics is one of the most important tools of medicine and public health in modern times, revealing the probabilistic causes involved in many kinds of diseases and disorders. It is true, though, that a simply mechanical sort of causation is useful for other health issues such as infectious diseases. The eradication of smallpox, for example, one of the great public health feats of the twentieth century, was made possible only because the specific microbe responsible for it was isolated so that a vaccine against it could be developed.

For other health problems such as noninfectious diseases, however, and for all sorts of epidemiological knowledge, other methods and other notions of causation are valid and highly effective. Skin cancer, we know from statistical studies, is caused in many cases by excessive exposure to sunlight. We commonly take steps against overexposure by using sunscreen and avoiding exposure during peak periods. In the summertime, our news programs even present us with the daily tanning index, which is really a sunlight exposure index. Although we do not know exactly the mechanism by which exposure is converted into cancerous cells, we do know that the exposure *causes* the cancer. Simply put, more exposure brings about more cancer.

This line of reasoning is so useful and so prevalent that it even has its own special name. One of the major figures in logical reasoning of the last century, John Stuart Mill, is famous for his four methods of logical inference. One of these is the method of "concomitant variation" by which an increase in an independent variable leads to an increase in a dependent variable, and a decrease yields a decrease. Even though the specific mechanism at work might not be known, we are justified in concluding that one causes the other.

Another example of statistical causation comes from public safety. Statistics plays a key role in understanding and dealing with death and injuries from motor vehicle accidents. We know that installing and using restraint systems such as seat belts and airbags results in fewer deaths and injuries from these accidents. We also know that, without these devices, deaths and injuries are more frequent. Furthermore, we know that we cannot predict exactly who will have a motor vehicle accident or when or how severe it will be because these concern the individual and many factors that are not precisely known and cannot be precisely controlled. We do know about the aggregate population, however, and can make statements about it with near certainty. And on that basis—statistics dealing with whole populations rather than with specific individuals—lawmakers feel sufficiently convinced to enact laws requiring the installation and use of restraint systems.

To keep our discussion from becoming too complicated, this chapter considers only smoking rather than other forms of tobacco use and only lung cancer rather than other cancers, diseases, and disorders. For over fifty years the statistical connection between smoking and cancer and other diseases has been known and generally accepted among scientists. Each year this knowledge is reinforced as more and more studies show the same statistical connection. Heavier smoking and over a longer period correlate with greater incidences of disease and shorter life expectancy. This is true regardless of gender, race, or nationality (the World Health Organization study mentioned earlier demonstrates this very clearly across the entire world). We also know that no other conceivable factor could explain the strength of the observed correlation other than smoking itself.

What is not known is the specific, microscopic mechanisms by which smoking causes cancer and other diseases. We do not know what genetic factors make one person more susceptible to cancer than another person. We do not know exactly how these genetic factors interact with environmental factors to bring about cancer. Although we do know that many specific chemicals in tobacco smoke can cause cancer, we do not know exactly how they interact among themselves or with other factors. But though we do not even know exactly how tobacco smoke changes healthy cells to cancerous ones, we do know that it does happen and very predictably. Nevertheless, despite not knowing the deterministic details of why one smoker would contract lung cancer whereas another smoker would not, practically all scientists and doctors studying smoking agree that smoking causes lung cancer as well as other diseases. There is a clear connection, and it is one of causation rather than coincidence.

One of the legendary events in the field of epidemiology is strikingly apt for the topic of smoking and cancer because it shows how a disease can be controlled even though the exact, microscopic agent involved is unknown. The nineteenth-century scientist Dr. John Snow was ahead of his time in many ways. One of these ways was to think that cholera is caused by some poison coming from people and transmitted from some humans to other humans through contaminated water. The dominant view at that time, however, was that poisonous vapors from the environment caused the disease. Keep in mind that this was well before the germ explanation of infectious diseases had become prevalent.

In 1854 in London, a series of serious cholera outbreaks occurred. Dr. Snow, thinking that the contagious agent was transmitted by contaminated water, studied the geographic and temporal patterns of the most recent outbreak. In a single neighborhood, 500 outbreaks occurred in only ten days. Snow noticed that the outbreak centered around a single water pump servicing the area. He advised public officials simply to remove the handle from the pump, forcing local citizens to obtain their water elsewhere. When the handle was removed, the local outbreak subsided. Keep in mind that Snow did not precisely know what in the water communicated the disease or how the agent worked its effects, but his statistically derived hunch was eminently successful. Snow's methodology and scientific insight stand as a landmark in epidemiology and public health.

For our purposes, causation is fundamental to examining the role of ethics in technical and scientific communications about smoking and disease.[2] How the term *cause* has been used in these communications clearly reflects the value system of the communicators and the goals they value and work toward. This matter is both very modern and very ancient. Though the question of what "cause" means is very modern, reflecting the most sophisticated thinking in contemporary science and philosophy, in some ways it is also very much like issues raised by the sophists mentioned earlier.

The sophists have been traditionally known as hustlers and charlatans, clever users of words to manipulate and deceive. As the modern analogue of this traditional view, the common stereotype of the car salesperson seems about on the mark. If you were to consider buying a certain expensive car, a salesperson might tell you that you would realize tremendous savings. In this way, spending becomes its opposite, saving, as you are seduced into reconceptualizing your finances. By fiddling around with the various interest rates, amortization periods, down payments, and trade-in allowances, the buyer is readily confused by the salesperson.

The sophists appeared at the very beginning of the study of rhetoric as the art of persuasion. They explored the power of language to alter beliefs and to shape the course of our actions. They offered to dispense the skills of this art to those willing to pay handsomely for them. They were interested in winning arguments, prevailing in debates, and achieving favorable judgments for oneself. Although it is true that in recent years this entirely negative view of the sophists has been challenged, the negative view of them has been generally accepted for nearly all of the past twenty-five centuries.

---

[2]Though the tobacco industry has insisted on calling the issue "smoking and health," most investigators feel that this terminology misleadingly links smoking with health. A clearer and more correct terminology is "smoking and disease and death." Some organizations opposed to smoking prefer "smoking *OR* health."

Newer thinking about the sophists, on the other hand, emphasizes their earnest exploration of the power of language to persuade but also to constitute meaning, knowledge, and values. It also emphasizes the threat posed by the sophists' powerful use of language as they challenged the prevailing social structure of power and control that excluded outsiders such as themselves. The sophists ended up not prevailing over their critics, and this explains, according to this view, why the sophists have been disparaged and excluded from a significant role in intellectual and social history—not the rightness or wrongness of their philosophy or methods but only the fact that they lost and were disempowered, the newer thinking explains. To the victors go the spoils, including the writing of history.

The sophists taught and practiced a base or disreputable sort of rhetoric, we are told by those who opposed them, such as Plato. Indeed, Plato is famous for his attacks on the sophists as manipulators who wanted only to win over their opponents and were willing to say anything to achieve that end. Often this involved just telling people what they want to hear and approving of what they want to do rather than telling the truth. If there was any "ethic" to their teaching, it was only that of winning and getting what one wanted. Plato and others such as Aristotle, however, felt that rhetoric—the power to persuade through discourse—should be based on and aimed toward ethical principles that seek what is good, right, and true. For these rhetoricians ethics comes before rhetoric.

The techniques of the sophists as traditionally represented do seem to closely resemble those of the tobacco industry, and our ethical judgment of them is similarly negative. The newer thinking, to be sure, undermines the traditional view of the sophists and celebrates them as innovators, intellectual explorers, and brave social critics. The sophists were noted for eristics, arguing for the sake of defeating an opponent (that is, rather than for the good of society). They also cultivated contention for the sake of contention, arguing of the sake of arguing. A good illustration of this frame of mind and value system is the famous assumption of Protagoras that there are two sides to any matter, and so any matter at all is always open to dispute. The sophists insisted on the two-sidedness of any issue and cultivated the ability to argue successfully on either side. For that reason, one of the more common charges against them was that they made the greater claim seem the smaller claim, and the better claim seem the worse. For them, everything and anything was debatable. And, of course, the aim of debate was for their own side to win. Their aim was not the choosing of the most ethical course of action.

How does this relate to the tobacco industry? From the very first serious medical and scientific claims that tobacco smoke causes or is causally related to cancer, the tobacco industry defended itself by going on the offensive. It aggressively sought out ways in which the statistical connection could be characterized as anything but causal; it solicited scientists and doctors who were willing to oppose the representation of the connection as causal; and it developed clever distractions from the focus on cancer. What is most important for our purposes is the way language was manipulated to suit a particular interest, and the paramount significance of the values lying behind and guiding that language use. The values and interests of the tobacco industry came before the language and gave shape to what was claimed as knowledge, not the other way around.

Let us take a look at an early document. In 1958 the British American Tobacco Co., Ltd., an enormous worldwide tobacco conglomerate that included the U.S. firm Brown & Williamson Tobacco sent investigators to report on the state of knowledge in the United States and Canada on smoking and disease. Regarding causation, they found the following:

> With one exception (H.S.N. Greene) [whose results the reporters dismissed] the individuals whom we met believed that smoking causes lung cancer if by "causation" we mean any chain of events which leads finally to lung cancer and which involves smoking as an indispensable link.(2) . . . The majority of individuals whom we met accepted that beyond all reasonable doubt cigarette smoke most probably acts as a direct though very weak carcinogen on the human lung. The opinion was given that in view of its chemical composition it would indeed be surprising if cigarette smoke were *not* carcinogenic.(6) . . . Research on smoking and health during the last few years has certainly convinced the majority of scientific opinion in the USA that cigarette smoke is capable of causing lung cancer in man, without defining exactly what is meant by "causation" except that cigarette smoke itself is an indispensable link in the chain of events leading to cancer.(8) . . . Although there remains some doubt as to the proportion of the total lung cancer mortality which can fairly be attributed to smoking, scientific opinion in the USA does not now seriously doubt that the statistical correlation is real and reflects a cause and effect relationship.

This was 1958, remember. Even then, the tobacco industry in the United States had already begun an aggressive campaign to discredit the scientific reports and to deceive the public. They created the Tobacco Industry Research Committee (TIRC) (which was later renamed the Council for Tobacco Research or CTR) to conduct research on possible smoking hazards. They also worked feverishly to craft documents that were critical of scientific reports and contested their validity. They also vigorously sought out any scientist or doctor who would endorse their position—and found a few.

How these few opposing voices were represented is what is interesting for our purposes. A very few people were represented as credible and respected voices in opposition to all the others, which was a gross distortion of their standing in the field and of the feelings of the field as a whole. More important, the bare existence of these differing voices was said to indicate that there was a real controversy about whether smoking causes cancer was taken seriously in scientific and medical circles. The true reality was that practically all scientists took these findings of causation seriously at face value. Recall the sophists' insistence that every topic has two sides worthy of argument, and their reputation for contention for the sake of contention and for the sake of their self-interest.

In this chapter we examine specific details from several significant documents in the monumental struggle between the tobacco industry and various public and private groups. This protracted struggle covers nearly fifty years, has cost hundreds of millions of dollars, and has generated millions of pages of documents. We can examine only a few of them, chosen to illustrate key points. Our focus is principally the concept of "cause," as well as other concepts. Though these concepts are largely scientific, we can also usefully consider them as technical because they have to do with specialized knowledge and because the application of technical concepts to practical affairs is a fundamental concern in nearly all technical communications. Some of these documents have to do with advertising. We can learn from them, too, because they, like most technical documents, concern not only the stark presentation of information but also the representation, interpretation, meaning, significance, and use of that information.

The immense volume of the interrelated documents can be overwhelming, making it difficult just to "get a handle" on them and to draw specific, meaningful, ethical lessons from them. For the sake of clarity, we will focus mostly on documents that have come to light from a single, recent landmark legal debate: the civil case of the State of Minnesota and Blue Cross and Blue Shield of Minnesota versus the major American tobacco producers (and one British producer) and the tobacco research group they created. As a result of this case, an enormous number of documents became publicly available that had previously been withheld by the industry from public examination.

In another chapter we considered the *Challenger* disaster and in this one the tobacco industry. Though they both have similar concerns with technical means and ethical issues, in a very important way they are quite different. In the case of *Challenger*, those who chose to overrule the engineers' concerns, to reinterpret factual information, and to take dangerous risks to achieve the organization's goals did seem to behave unethically. They probably did not truly believe, however, that it was very likely that the *Challenger* astronauts would die as a result of their decisions. They were not certain that the mission would end in disaster, as they had asked engineers to try to prove, which they could not do.

In the case of the tobacco industry, however, the knowledge of harmful effects was known with practical certainty, regardless of the industry's denial of such knowledge. This after all is the basis of the many governmental lawsuits against the industry. They knew that substantially increased cancer, heart and circulatory disease, and respiratory disease would definitely occur to a substantial number of people, numbering in the millions over the years. They also knew of the substantial decrease in life expectancy for nearly all chronic smokers. They also knew that smoking is addictive and that therefore consumers constituted a captive and often unwilling market. Therefore they knowingly and willfully brought certain disease and death to a portion of their consumers and did so over the course of several decades in direct defiance of the urgings of governmental and medical authorities. They were also conscious of their continuous deception as they schemed for language and arguments to avoid any public acknowledgment of what they well knew. For this reason, the tobacco decision makers behaved much more unethically than the *Challenger* decision makers.

DOCUMENTS

The entire U.S. tobacco industry has been under intense scrutiny and criticism in recent years, leading to moral outrage and to a number of major lawsuits by individuals and governmental entities. The tobacco industry in 1997 negotiated some of the terms of an out-of-court settlement of gigantic magnitude, on the order of 350 billion dollars over a period of years. This amount is much greater than the gross national product of most of the world's countries. That settlement now is under its own intense criticism partly because it would seem to bar further criminal and civil action in the courts. Not only are some governmental entities opposed to barring further action but the industry itself has objected to some proposed amendments to the settlement.

We should keep in mind as we examine these documents that part of the context of this suggested settlement is that the tobacco industry had never, until very recently, lost a case in the courts over the damaging health effects of its products, their addictiveness, or their advertising methods. On the face of it, this fact might appear to suggest that they were in the right and the plaintiffs in the wrong, but, like so much of the information about the tobacco industry, appearances are deceiving.

The fuller, truer reality is that the context is complicated in many inapparent ways. One has to do with the sophistical treatment of causation discussed earlier, by which the most extreme, impossible standards for causation are insisted upon by the industry solely for the sake of their self-interest. Another has to do with the industry's ability to identify even a few legitimate scientists who would quibble on the issue of causation. Still another way is the tremendous profitability of the industry, which gives them enormous financial and legal resources to use on their behalf as well as an indefinite expanse of time. Most claimants against them, however, have few such resources and only a limited time. All too often, the claimants died from tobacco-caused diseases before the jury deliberated. This profitability also yields enormous tax resources to the states involved, leading to support for the industry at high levels in their state legislatures. The most important way that the apparent state of affairs differs from the true reality is that, every time a case has approached a culmination that might yield a decision unfavorable to the industry, the industry has negotiated an out-of-court settlement. Thus there are many cases that they really "lost" in terms of rightness and money, but none that carry the weight of establishing a legal precedent. Above all, the industry had avoided not only a single judgment against them but had also avoided the crucial *first* judgment against them, which would act as a watershed yielding a cascade of further cases based on that precedent.

These judicial matters have yet to be completely resolved. The charges are very serious, including fraud, conspiracy, negligence, false advertising, and product liability, and are of an unprecedented magnitude. Because of the absence of firm judicial determinations, much of the public discourse about the industry and the continuing harm that it has brought to the public has been couched in tentative, suggestive language rather than direct, firm statements. The reason is to avoid lawsuits from the industry.

Before turning to the language and arguments of specific documents, we should note another dimension of the context that might not be apparent. The legal system in the past few years has repeatedly tried to build a case, which it feels is warranted, against the industry. The mass of material that could serve as potential evidence has been enormous, tens of millions of pages of documents. These documents, though, come from an industry known to be highly secretive, closed, and guarded about its privacy. For this reason, many of the documents we have to work with are worded in obtuse, contorted ways to disguise their direct substance.

The documents also reflect an intense effort, which has increased dramatically over the years, by the legal staffs of the tobacco industry to ensure that these documents could not serve the interests of potential claimants. They made sure that all discussions about health research and about policy decisions were made in the presence of lawyers so that they could be hidden from discovery under the rules of lawyer-client privilege. In addition, since about the mid-1970s, the industry has sought to insulate itself from what it perceives as potentially problematic information; it took active steps to keep itself ignorant of and insulated from what it did not want to know.

Finally since about the mid-1970s, the industry has taken active steps to distance itself from the information it already had. Files were purged and potentially incriminating documents were shredded. The remaining documents were not indexed on written lists that could be discovered by claimants but were kept track of only through personal, verbal knowledge. Simply put, the industry tried to conceal or destroy any potential "smoking guns." Nevertheless, enough written material exists for many states to form strong legal cases against the industry.

We will analyze several important documents from the tobacco industry over a period ranging from the early 1950s to the late 1990s. We will see how the industry exaggerated, distorted, or misrepresented information in order to mislead the public. We will see too how these documents illustrate a sophistic approach to language, logic, and reality, in effect trying "to make the worse seem the better" and trying to establish two differing sides to an essentially single-sided matter. (Keep in mind that there are other legitimate meanings of the term "sophistic" that have emerged in recent years.) On the face of them, the documents appear to be reasonable, cautious, and careful as long as one does not examine their claims critically and without consideration of their overall context. We will see not only how language and argument can be used to craft misleading documents but also see the powerful role of the values (or lack thereof) lying behind the text of a document.

1950s

In the early 1950s several medical research reports were published linking smoking with lung cancer, other cancers, and other diseases. Among the more prominent of these was that of Dr. Ernst Wynder published in the *Journal of the American Medical Association*. In addition, the American Cancer Society had become so concerned that it conducted a research study of unprecedented magnitude: a long-term study over many years with 22,000 volunteers in eleven states.

In response to these reports and the concern they evoked among the public, chief executives of all but one of the major tobacco companies met in December 1953 to mount a defensive strategy. They concluded that they needed to launch a public relations campaign that would be "entirely 'pro-cigarette' in nature." The public relations firm of Hill and Knowlton was enlisted for this campaign. The Hill and Knowlton memorandum of December 24, 1953, is notable for its basic stance toward the medical studies: "These developments [the medical studies] have confronted the industry with a serious problem of public relations" (1). The problem, we should note, is not one of concern for health but for *public relations*. Their plan was to form a research group supposedly independent of the tobacco industry (which was anything but independent, as we will see) in order to get at the "real" facts of the matter. Though on the face of it, the desire to "get the real facts" can hardly be faulted, the fuller reality is that "the facts" would be limited to only those facts that opposed the emerging medical findings. Their aim was to complicate the matter generally and to divert attention from smoking per se to other causes of cancer and disease.

This key memorandum begins the long train of counterargument about absolute cause and proof. It states the need to reassure the public that "there is no proof that cigarette smoking is a cause of lung cancer" and that "there is no proof to the claims which link smoking and lung cancer" (2, 4). Although technically it might be true that at the time the causal connection was not absolutely known, the ethical issue was how to represent this lack of absoluteness. Does the industry choose to see it as showing the connection to be untrue, or does it choose to see it as strongly suggesting a very real health hazard? The fundamental attitude—and the implicit value system—here is vitally important. One version looks out first and foremost for the industry, whereas the other puts the public health first. It is a question of choice because that is what ethics is really all about, making conscious and conscientious choices, which are then expressed in language.

The result was a lengthy textual advertisement titled "A Frank Statement to Cigarette Smokers" published in January 1954 in 449 major newspapers throughout the country. (At that time, newspapers were the primary source of news and information for nearly everyone, much more so than TV.) It presents the position that the tobacco industry will argue from for the next forty years, namely that the terms "cause" and "proof" cannot accurately be used regarding smoking and disease; that the scientific and medical community is divided on the matter; and that the industry believes there is no health danger from smoking. All of these points are ethically questionable because they are equivocations and evasions of the reality of the research.

The statement has two sections. The first section addresses recent reports about the dangers to health of cigarette smoking and attempts to discount them. The second states that the tobacco industry feels a great responsibility for the health of the smoking public and presents a plan of action to discharge this responsibility by forming an industry group to research the health effects of smoking. This group was called the Tobacco Industry Research Committee (TIRC), which in 1964 was renamed the Council for Tobacco Research (CTR).

This statement is one of the most famous and widely cited documents by the industry on the health effects of smoking for three reasons:

1. It articulates a position central to the industry's subsequent justifications of its activities, namely that causation and proof have not been demonstrated.
2. It shows the frank disdain by the industry of scientific, medical, and technical research accepted by nearly all medical and scientific professionals outside the industry.
3. It attempts to lay upon industry a mantle of scientific honesty and rigor that has been found to be false and deliberately misleading.

It is a technical document because it communicates complex technical information, specifically the results of medical research, to the non-technical public. In the process, it represents the industry as a protector of the public health and an authority on scientific rigor. Though it interprets technical information only from their point of view, it does not say this directly. Instead it undermines the conventional wisdom about medical research reports, namely that they tell the truth. It also appears to seek to educate the reader and to explain this complex technical information in a way usable to the audience.

The document begins with careful wording meant to leave a misleading impression. It refers to recent reports linking smoking to cancer as "experiments" associated with a "theory." While both these terms are not strictly incorrect in this case, their connotations for the lay public would cast these reports as speculative and not factual. Furthermore, these reports, it says, have given "wide publicity" to this theory, which has the connotation of a false appearance meant for show but without much true substance. This makes the scientific reports seem watered down and indefinite while it represents the matter as one concerning only public impressions rather than scientific knowledge about health and disease. This is consistent with the industry's basic approach of seeing the research reports as a public relations issue. The whole effect is to make a statement that could be defended as technically correct but which also would leave the misleading impression that the medical and scientific reports are only conjectures.

The document also portrays the industry as a victim of malicious publicity from other agents, while in reality the public is the victim of the industry. This turning of the tables, a reversal of conventional thinking that flies in the face of common sense, is exactly one of the traditional meanings of the term sophistic. The sophists repeatedly made use of this sort of thinking and speaking, taught it, and were famous or infamous for it, depending on your perspective. Gorgias's famous "Economium of Helen," for example, opposed the traditional view of Helen of Troy as an unfaithful wife who deliberately abandoned her spouse, by arguing that she was only a helpless victim of her emotions and of the powerfully seductive words of Paris. The victimage is completely reversed in "Encomium of Helen" just as it is in "A Frank Statement." Now Helen is the victim rather than her husband Menelaus. Notice also the tone supporting victimage even of the title "A Frank Statement," which suggests that it speaks the truth as it opposes the supposedly less frank and supposedly less true statements made by its opponents.

A similar example of this sophistic arguing from appearances and claims rather than true reality comes from a later 1970 memorandum from Helmut Wakeham, the director of research at Philip Morris, to the company's president, Joseph Cullman. In reference to the Council for Tobacco Research, successor to TIRC, the supposedly neutral third party that would get at the "real facts," Wakeham says:

> It has been stated that CTR is a program to find out "the truth about smoking and health." What is truth is false to another. CTR and the Industry have publicly and frequently denied what others find as "truth." Let's face it. We are interested in evidence which we believe denies the allegations that cigarette smoking causes diseases (Minnesota hearings of July 16–8, 1997, item LG 0209296).

Returning to "A Frank Statement," the second sentence says that the experiments on which these reports are based are "not regarded as conclusive" among researchers in this area. Unfortunately, though this statement too can be seen as technically correct in an extreme, absolute sense, by design it will be read by most of the audience as showing that the research means little.

In support of its position, the statement list four points made by "distinguished authorities" meant to undermine the credibility of any causal connection between smoking and cancer. The first point states that lung cancer has many possible causes, a true statement but which sidesteps the high prominence of smoking among these causes. The second point states that there is "no agreement among the authorities" as to what the cause of lung cancer is. While it is true in an absolute sense that 100 percent of authorities do not agree as to the cause, it would take only a single person among thousands of researchers and doctors to make this statement true. By leaving out the technical detail of the numerical proportions involved, it leaves the suggestion among most casual readers that disagreement is widespread, which was precisely not the case. The third point states that there exists "no proof that cigarette smoking is one of the causes." This is an instance of the industry insisting on unrealistic standards for certainty that it did not apply in its own research. Furthermore, in being addressed to a non-technical audience, this statement allows the commonsense, lay meaning of the word "proof" to misleadingly resonate for the reader, though they could insist that their true, intended meaning was only the stringent standard of absolute certainty. The fourth point contends that conclusions drawn from statistics can be misleading or confusing, and that, furthermore, the "validity" of the statistics "is questioned" by many scientists. It does not explore what it means by validity or why it might be questionable, again leaving the reader with a suggestive impression without having to open up the basis for this claim to public examination. It also fails to mention that the same methodologies had been and continued to be used by the industry in its own research, indicating that for the most part they themselves accepted the validity of these methodologies. The overall impression was to undermine the scientific credibility of the research reports, though the research was then and is now accepted as sound, while highly overstating the disagreement among authorities. Central to this misrepresentation was developing the appearance of the industry conscientiously insisting on technical accuracy, an appearance exactly opposite to the reality.

After impugning the conclusions of research authorities accepted within their fields in the first section, the second section of the statement proceeded to try to sanctify the industry by claiming its great concern for the public health and its eagerness to take active steps to protect it. It states that concern for public health is a "basic responsibility, paramount to every other consideration" in the tobacco industry. The statement goes on to claim that the industry's belief is that their products are "not injurious" to the public. As we will see later in testimony to congressional hearings in the 1990s, a belief per se is not the same as knowing and is almost impossible to refute.

Also in this section, the industry (represented by all but one of the major U.S. tobacco firms as sponsors of this statement) pledges to aid in research through the Tobacco Industry Research Council which will consist of distinguished authorities in medicine and science and be headed by a scientist of national reputation. While these qualifications of its members held true, at least initially, what is not stated is whether these authorities are representative of the opinions and thinking of their fields. From the start, the voices on the council were unrepresentative and became even more so as time went on, to the point of absurdity.

For decades afterward, the industry continued to say in public print that there was a real "controversy" about whether smoking causes lung cancer. Its internal documents, however, show these statements to be lies and gross distortions. In the 1958 memorandum cited earlier, for example, the huge international tobacco company British and American Tobacco Co., Ltd. (BAT), sent representatives to gather a sense of the causal connection within the U.S. tobacco industry. Its findings are clear and unequivocal:

> With one exception (H. S. N. Greene) the individuals [managers and researchers] whom we met believed that smoking causes lung cancer if by "causation" we mean any chain of events which leads finally to lung cancer and which involves smoking as an indispensable link. . . . Greene of Yale still says that his repeated failure to produce carcinoma [cancer] by implanting lung tissue along with tobacco smoke condensate into the muscles of mice is conclusive evidence that smoke cannot cause lung cancer. His experiments were not done quantitatively, however, and on these grounds alone the conclusion which he draws is certainly not justified. We found disagreement however as to the likely *mechanism* by which smoking may cause lung cancer. (2)

It concludes with this statement:

> Although there is some doubt as to the proportion of the total lung cancer mortality which can fairly be attributed to smoking, scientific opinion in the U.S.A. does not now seriously doubt that the statistical correlation is real and reflects a cause and effect relationship. (8)

Over the years following, the evidence of the cause-effect relationship would continue to accumulate into an unequivocal mountainous mass pointing in only one direction. Nevertheless, the industry would continue to claim that there was a real controversy in the scientific community about the causal connection and that more research needed to be done before anything certain could be said—and, more important, before they would feel the need to take any corrective action. The industry even formed (as well as sponsored and ruled) the Tobacco Industry Research Committee (later named the Council for Tobacco Research) to find the "true" facts about the matter.

Notice the intentions reflected in these documents, the ambiguity of their language, and their deliberate distortion and misrepresentation of technical information. These documents are forms of technical communication because all the information they speak of is scientific and technical, and the arguments they make are of a scientific and technical nature. What do we mean by "cause" and "proof," for example? What do the data really mean? What methodologies would yield convincing findings? How much evidence would it take to be persuasive? What action should be taken in light of the available evidence? Such questions arise continually in nearly all complex technical communication situations, especially in technical reports.

If we feel that these tobacco industry documents are ethically questionable, there is a good deal of reputable support for our judgment. Dr. C. Everett

Koop, former Surgeon General, Dr. David C. Kessler, former head of the Food and Drug Administration, and Dr. George D. Lundberg, for example, in an editorial in the prestigious *Journal of the American Medical Association* state their judgment very directly:

> For years, the tobacco industry has marketed products that it knew caused serious disease and death. Yet, it intentionally hid this truth from the public, carried out a deceitful campaign designed to undermine the public's appreciation of these risks, and marketed its addictive products to children. The industry long ago knew that nicotine was addictive but kept its findings secret and consistently denied the fact, even as overwhelming evidence to the contrary eventually emerged. By these actions, the tobacco makers have shown themselves to be a rogue industry, unwilling to abide by ordinary ethical business rules and social standards. . . . The tobacco industry has intentionally designed and marketed lethal products and deliberately hidden their well-known risks. These actions are morally reprehensible. (50)

The disdain of scientific and medical research in "A Frank Statement" spilled over to both formal statements and popular advertisements by the tobacco industry. An advertisement for Old Gold cigarettes on the same date as "A Frank Statement," January 4, 1954, illustrates the industry's willingness to make light of the research while manipulating popular opinion in the same direction. The dramatic, two-page spread consists mostly of an image of the cigarette package with a short textual message. It expresses their appreciation to American cigarette consumers for increased sales despite emerging reports of the hazards of smoking, thanking them specifically "for putting your trust in the cigarette made by tobacco men . . . not medicine men (Old Gold, *Life*, January 4, 1954, 36/1, pp. 50–51). Notice that this statement makes no explicit claim about the scientific validity of recent medical reports. It focuses simply on where consumers choose to place their trust, but does not address whether that trust is warranted. Clever word plays such as this flourished in popular advertisement for decades afterward. These statements with mixed serious and comic meanings in effect allowed the industry to have it both ways. The industry can simultaneously recognize the growing concern about technical information in the form of medical research reports, while deflating this concern through mocking humor.

1960s

In the early 1960s the U.S. Surgeon General appointed an advisory committee to investigate the health effects of smoking. The industry had to respond to governmental concerns from such a high level. In a memorandum from Brown & Williamson Tobacco Co. (B&W) marked "STRICTLY PRIVATE AND CONFIDENTIAL," a lawyer for the company stated:

> Whatever qualifications we may assert to minimize the impact of the [Surgeon General's] report, we must face the fact that a responsible and qualified group of

previously non-committed scientists and medical authorities have spoken. One would suppose we would not repeat Dr. Little's oft reiterated "not proven." (1)[3]

By this time the industry could find very few reputable scientists willing to defend their position. They had to call on the same few people over and over to claim that a causal connection was "not proven." These few voices against an overwhelming chorus of highly respected voices came to sound ridiculous and unbelievable.

When the Surgeon General's first official report on smoking was published on January 11, 1964, in a 378-page document, Kluger explains, it "had to omit certain judgments that a majority but not all of its members would have endorsed. Every nuanced word was reviewed scrupulously, assuring the most cautious possible language consistent with the evidence" (255). It did, however, go so far as to say, "Cigarette smoking is causally related to lung cancer in men."

There had been clear agreement among the panel that smoking was a serious health concern but disagreement as to whether the evidence warranted using the term "cause." Dr. Charles LeMaistre had reservations about strict demonstration of causation, Dr. Walter Burdette wondered about additional factors such as heredity that might complicate the picture of causality, and the director, Dr. Hammond, had the ear of Dr. Little. "What emerged," says Kluger, "was a highly detailed, closely reasoned, but far from combative report, that was substantially compromised. Understated and embodying the lowest common denominator of agreement among them," the report pulled its punch. It concluded only this: "Cigarette smoking is a health hazard of sufficient importance in the United States to warrant appropriate remedial action."

By this time, too, unimpeachable evidence that nicotine is addictive had been found. The prestigious Battelle Institute for medical research was enlisted by the industry to investigate the mechanism of this addiction. The Battelle Institute developed one hypothesis to "explain the addiction of the individual to nicotine," which it took as already proven, and stated outright: "Nicotine is addictive. We are then, in the business of selling nicotine, an addictive drug." (4). Interesting here is the line of inference. The addictiveness of nicotine was taken as accepted even though the exact mechanism by which the addiction occurs or operates is not known. Note that the industry was unwilling to embrace exactly the same line of inference regarding smoking and lung cancer— it assumed elsewhere that if the exact mechanism is not known, then the causal connection cannot be assumed. Here, on the other hand, the causal connection is accepted even though the exact mechanism is not known.

---

[3]Dr. Clarence Cook "Pete" Little was originally a moderate opponent of smoking as the first head of what later became the American Cancer Society. When he was replaced as head by a more vigorous and more socially sophisticated nonscientist, he became a critic of the emerging research. His objections were in part responsible for the first Surgeon General's report being so mild and qualified on the causal connection.

Also contained in this memorandum is a truly frank (and secret) statement about the real purpose of the Tobacco Institute Research Committee (TIRC). The industry had claimed publicly over and over that this committee, which it created and supported, was totally impartial and free to pursue whatever research it felt was needed and that its Scientific Advisory Board (SAB) of eminent scientists was totally disinterested. This document, however, states that its true purpose was not disinterested science but public relations propaganda intended to oppose genuine scientific research. After calling for truly scientific, disinterested research into the connection between smoking and cancer and cardiovascular disorders, the writer states:

> The TIRC cannot, in my opinion, provide the vehicle for such research. It was conceived as a public relations gesture and (however undefiled the Scientific Advisory Board and its grants may be) it has functioned as a public relations operation. Moreover its organization, certainly in its present form, does not allow the breadth of research—cancer, emphysema, cardiovascular disorders, etc.—essential to the protection of the tobacco industry. (2)

It could not, that is, conduct research that would win genuine respect in the scientific community. Notice that the principal interest of the writer and the industry here is to protect the industry, not to protect the health of the public. In this way, values strongly guided what technical information was reported and how it was represented.

The concern by the industry with self-protection above all else—including above concern for the health of the public—is underscored in a long memorandum at a major international conference of the tobacco industry in Southampton in 1962. The minutes of the meeting report: "Mr. Reid [W. V. Reid of the B and W Australian affiliate] suggested that no industry was going to accept that its product was toxic or even believe it to be so, and naturally when the health question was first raised, we had to start by denying it at the P.R. [public relations] level" (Minnesota Doc. ID 1102.01, p. 45, Jan. 1, 1962).

1970s

By the 1970s the industry had become highly interested in filtered cigarettes as a way to respond to increasing health concerns among the public and in government. They found themselves entrapped in their own web of deceit, however, in trying to justify and market the newly developed filtered cigarettes. If filters are supposed to be desirable because they filter out materials that cause health problems, then of course one must assume and acknowledge that smoking causes health problems, which, however, the industry had been strenuously denying. The tactic they settled on was to try to justify the filtering of cigarettes on the basis of only a *perception* among the public that smoking is linked to health problems, without actually acknowledging the reality that smoking causes health problems.

One way the industry carried out this distorted misrepresentation of the medical evidence, essentially technical information, was to use euphemisms. In a July 20, 1970, memorandum on the direction of research and development activities in the industry, held at Chelwood, England, D. J. Wood of BAT research explains:

> You will remember that it can be shown experimentally that smoke, or smoke condensate, has certain undesirable effects on animals, including tumor production on the backs of mice. At R & D. E. [research, development, and engineering] we cannot disregard these results, although in themselves they do not prove any connection between smoking and human disease. It is our belief that the cigarette of the future must have reduced biological activity, and when I speak of biological activity I mean those adverse effects such as tumor production on mice. (2)

The report goes on to outline research plans for ways to reduce "biological activity," a euphemism for cancerous tumors. Notice too that the document tries to have it both ways, acknowledging a health hazard while denying it in the next sentence.

By now the public had become concerned about the damage not only to the health of smokers but also to the health of nonsmokers through secondhand smoke, what the industry preferred to call "public smoking." The public and the government were moving toward banning smoking in various public areas such as government offices and transportation vehicles. In its white paper on this topic on April 18, 1979, the industry began by reiterating its sophistic claims that there are two sides to the matter and that a controversy exists, though by this time their counterclaims were becoming ever less believable. The paper begins:

> The tobacco industry acknowledges that there is controversy over many aspects of the general problem of smoking and health. There is disagreement among medical experts as to whether the reported associations between smoking and various diseases are causal or not. . . . It is self-evident that, since there is disagreement among medical experts as to whether the reported associations between smoking and various diseases are causal or not, there is no compelling evidence amounting to proof that there is any causal link between "Public Smoking" and various diseases. (1)

Notice the diversion of focus in the first sentence. "The tobacco industry acknowledges" suggests that they will finally admit what we and everyone already knows. But the sentence goes on to develop instead a focus not on knowledge or facts but on "controversy." The notion of acknowledgement thus is inverted to refer not to the publicly accepted connection between smoking and disease but to the suggestion that the public, too, should now acknowledge the supposedly obvious, that a controversy even exists. Thus the suggested focus on "smoking and health" is turned on its head to "controversy" instead. Inversions of meaning and opposition to generally accepted knowledge are hallmarks of sophistical argumentation. And this is only at the level of word choice.

When we try to find out what exactly this language refers to, we find no correspondence to reality. Basically, what the writer says actually has no connection to reality. Recall that critics of the sophists typically accused them of leading their audiences to mistake claims for truths. In this case, the writer repeatedly *claims* that controversy and disagreement exist (twice in the first page and several times later as well) even though none exists in reality. Practically no reputable scientists doubted that the correlation was causal, though some scrupulous scientists might resist using the term "cause" unless the actual mechanism of causation were also known. No scientists outside the industry doubted that eliminating smoking would eliminate a great deal of lung cancer and other diseases. Presumably for this writer, however, the repeated statement of the claim was meant to reinforce its believability in the audience.

This same 1979 document also shows another sophistic twist to the industry's argument. Among the reasons that the industry needs to take further action is: "Claims that tobacco smoke in the atmosphere causes disease in nonsmokers are unsubstantiated; nevertheless, a large portion of the population—both smokers and nonsmokers—believe these claims" (2). By using the term "believe," the writer can sidestep the question of whether theses beliefs are well founded and warranted or, on the other hand, unfounded and unwarranted. It deliberately distances the belief as well as the focus of attention away from factual evidence. This deceptive, self-serving use of the term and notion of "belief" will rise to prominence again in 1994 when the chief executives of all but one of the major tobacco companies swore under oath that they did not "believe" that nicotine is addictive.

By the mid-1970s, the industry's own research had become so clear and compelling about the health damages caused by smoking that they closed down several of their research operations in order to prevent discovering information they did not want to learn. In the widely cited Pepples memorandum of March 10, 1977, the industry's own management and lawyers stated, "these tests are so-called red light tests," meaning that in the field of medical research such results are clear indications of serious health hazards, regardless of the particular mechanisms at work. The memorandum goes on to explain, "I do not have to tell you what Senator Kennedy [a prominent opponent of smoking] would do with a finding of red lights in one of these tests" (Minnesota Doc. ID 1910.02, 1).

From this point on, it was decided that lawyers were to have decisive control over which research proposals would be approved, what research would be terminated, how reports were written, and which reports would be published. They controlled TIRC, by then renamed the Council for Tobacco Research (CTR), even more than before. Potentially damaging research either was not funded or was redirected to be under the management of the lawyers. By doing this, it was hoped that the findings could then be kept from disclosure to the government and to the public under attorney–client privilege. In this way the public and the government could be kept in the dark, and the industry could continue to claim ignorance of any health hazards from smoking. The industry could have its cake and eat it, too, being able to contend that it was conducting active research while also being able to claim that research showing health hazards was inconclusive. "The point here is the value of having CTR doing work in a [seemingly] nondirected and independent fashion as contrasted with work either in-house or under B&W [Brown & Williamson Tobacco Co.] contract which, if it goes wrong can become the smoking pistol in a lawsuit" (1).

1980s

Moving to the 1980s, the control of information by lawyers in order to prevent disclosures that would be detrimental to the industry (though beneficial to the public) was tightened. A June 12, 1984, memorandum, for example, by the legal counsel of the British and American Tobacco Co., Ltd., an international tobacco conglomerate, reported on an ongoing case against the industry:

> Developments [research reports] have rendered product liability actions against tobacco manufacturers more difficult to defend in the 1980's and that adverse evidence which could be attributed to the defendants [tobacco companies] is a serious problem. . . . Direct lawyer involvement is needed in all BAT activities [including research] pertaining to smoking and health from conception through every step of the activity (1).

In dealing with research already in progress or information that had already developed, the lawyers struggled to find a way to make the industry's activities appear defensible. The memorandum continued:

> The problem posed by BAT scientists and frequently used consultants who believe [that] cause is proven [i.e., smoking causes disease] is difficult. A sound recommendation must be based upon consideration of several factors including the basis upon which senior management relies in concluding that the opinion of the scientists is incorrect; the overall reliance of senior management on opinions of the scientist; and the company's duty to encourage scientific inquiry. (2)

Though this statement sketches how a defensive stance could be developed, the memorandum does not offer any specific data or arguments to use in particular cases. The intent here is to reinterpret or re-represent a scientific finding in such a way that would be less damaging to the industry. In this way, scientific—and presumably technical—information could be controlled and distorted in ways that negate or dilute their scientific validity. Thus a scientific report can be transmuted into a truly unscientific report that nonetheless appears scientific. We should notice from this example that similar distortions could occur in many technical communication situations.

A few years later, the need to control information became even more desperate. In a February 17, 1986, memorandum, corporate counsel for B&W, J. K. Wells III, went beyond controlling how information was portrayed to controlling the entire reporting process and whole documents. Rather than receiving lengthy, fully explained documents, reports on research would now be limited to snippets in order to limit the revelation of potentially damaging information:

> B & W will receive concise reports, estimated to be about one-half page in length, twice each year for each project it wishes to follow. While the brevity of the reports will reduce the potential for receipt by B & W of information useful to a plaintiff, disadvantageous information could [still] be included and the reports could serve as road maps for a plaintiff's lawyer.

The memorandum goes on to explain how B&W in America can isolate itself from information about health hazards from smoking. Now they had not only to limit their own research and control the research done by TIRC/CTR but also had to limit information sent to them from other industry sources, including BAT International.

> I have advised that we can receive reports from some of these projects notwithstanding the risk. The response is that we cannot shut out the flow of information: the BAT will find ways to get information into B&W from the scientific projects it is running in its laboratories worldwide. The only way BAT can avoid having information useful to plaintiff found at B and W is to obtain good legal counsel and cease producing information in Canada, Germany, Brazil and other places that is helpful to plaintiffs. (1)

The memorandum ends with an attachment listing specific research projects that should not be approved or funded—basically information B and W does not want to know about. Just a year earlier, documents show, J. K. Wells of B and W also sent instructions for the purging of potentially incriminating files, another form of unethical information control.

These events and documents show how information was controlled in unethical ways. What research is funded and what is not funded naturally controls the information that would follow from such research, which was precisely the point of lawyers having the final say in such decisions. Likewise, if an ongoing research program seems to be leading toward final results that would be troublesome for the industry, the program is simply shut off, and so the potential results are never actually reported. In other cases in which results were actually achieved, the industry through its lawyers could bar the information from being published and reported to the public or to the government.

The language and the argument of a report could also be reworked by lawyers and managers to water them down, make them seem more tentative, or otherwise obfuscate the full meaning and significance of the information. Notice that even the standard closing section of a scientific report, in which the need for further research is explained, can easily be exaggerated to make the findings appear to be inconclusive.

Further evidence of the control of information was revealed in the PBS program *Frontline*. In March 1998 *Frontline* interviewed Dr. Gary Huber, a long-time smoking researcher who decided to become a whistleblower after learning of how the tobacco industry was controlling, distorting, and misrepresenting scientific findings. The industry "had used him, manipulated his research, and wasted his career," Huber told *Frontline*. "Every scientist that the industry had any contact with had a keeper, within the industry, a lawyer" who oversaw the research and any reporting about it, he said. The report explains that a lawyer was always present when Huber spoke with any tobacco executives in order to try to make his comments confidential under supposed attorney–client privilege (*Frontline* Web site, "Inside the Tobacco Deal," 17). More evidence comes from Jeffrey Wigand, a former vice-president of research for Brown & Williamson until he was fired in March 1993. His testimony to federal and state prosecutors has been vital, revealing the hidden workings of the tobacco industry and its documents. "Wigand testified that Brown & Williamson edited out sections of documents that might prove damaging in future litigation," *Frontline* states (*Frontline*, "Settlement Case," 12).

1990s

Moving to the 1990s, we see that public and private outcries, governmental inquiries, legislative actions, and judicial suits are finally beginning to take their toll. Huge volumes of documents that before had been kept from the public eye have come to light, thanks to whistleblowers and disenchanted insiders. In several states the industry has been unsuccessful in having documents barred from examination under the claim of attorney–client privilege. Several states have successfully sued the industry to recover the enormous costs to them for the medical care of those afflicted with smoking-caused diseases. In addition, a single major tobacco company, the Liggett Group (formerly Liggett and Myers Tobacco Company), has finally admitted that nicotine is addictive. In 1997, facing the possibility of astronomical judgments against the industry from innumerable claimants, the major tobacco companies had negotiated the terms of a single enormous settlement that would quell some of the public clamor. This settlement has run into delay and opposition, however, and the other tobacco companies are distancing themselves from it.

A SINGLE WORD

What's in a word? More specifically, what are the ethical implications of a single word? The PBS television program *Frontline* reported in 1998 on the significance of a single word in a communication from the tobacco industry. On April 14, 1994, the chief executive officers of all major U.S. tobacco companies testified under oath in person before a congressional committee that they did not "believe" that nicotine was or is addictive. The statement of Andrew H. Tisch, chairman and CEO of Lorillard Tobacco, is typical: "I believe that nicotine is not addictive."

At the time, most of the public and congress were outraged at what they took to be a flat lie under oath, though the documentary proof that the executives were lying was lacking. Not until May 12, 1994, with the mysterious appearance of a large box of documents at the doorstep of Prof. Glantz, the dedicated opponent of smoking, did documentary evidence come to the attention of the federal government. Thousands of internal reports, letters, and memoranda from B and W and BAT were in the box. After much wrangling between the tobacco companies, Prof. Glantz, and UCSF, the California Supreme Court determined that the documents could be released to the public. They are now housed physically and electronically at the UCSF library.

Upon learning of these documents from within the tobacco industry itself, especially those clearly showing that nicotine is addictive, the federal Department of Justice launched an investigation into possible perjury charges against those CEOs. Perjury seemed to have occurred because the documents showed the industry's management *knew* that nicotine is addictive. As of the middle of 1998, however, the Department of Justice has decided not to pursue the charge of perjury because of that single word, *believe*. As *Frontline* explained, the perjury charge "has been abandoned because of the difficulty of prosecuting anyone for their 'belief'" (2).

It does not take a great deal of imagination to think of instances in ordinary technical communication situations in which similar misrepresentations could be foisted on an audience quite unethically by a communicator. Almost all technical reports, for example, must draw an inference or conclusion from technical data. Though the data themselves might be incontrovertible, the drawing of an inference often involves a good deal of judgment in moving from concrete information to interpretation and application. In the case of *Challenger*, for example, the measured numbers indicating the depth, extent, and location of charring of the O-ring seals could have been redone with even more precision. The exact numbers were almost beside the point, however, because just the fact that *any* charring occurred was what was really important. In such cases, as in many other real situations, the numbers themselves can be of less importance than the interpretation that explains what they mean. And this almost always involves a judgment, both interpretive and ethical.

Some tobacco executives continue to quibble about the term "cause" in a self-protective gesture that has always minimized if not neglected the public health. On January 29, 1998, for instance, Geoffrey C. Bible, the chairman and CEO of Philip Morris Companies, Inc., made the following involved statement by the U. S. House Commerce Committee:

> We recognize that there is a substantial body of evidence which supports the judgment that cigarette smoking plays a causal role in the development of lung cancer

and other diseases in smokers. We previously have acknowledged that the strong statistical association between smoking and certain diseases, such as lung cancer and emphysema, establishes that smoking is a risk factor for and, in fact, may be a cause of those diseases. For example, of all the risk factors for lung cancer that have been identified, none is more strongly associated with the disease, or carries a greater risk, than cigarette smoking; a far greater number of smokers than non-smokers develop lung cancer. Despite the differences that may exist between our views and those of the public health community, in order to ensure that there will be a single, consistent public health message on this issue, we will refrain from debating the issue other than as necessary to defend ourselves and our opinions in the courts and other fora in which we are required to do so. For that reason, we are also prepared to defer to the judgment of public health authorities as to what health warning messages will best serve the public interest, as reflected in the proposed new health warnings. (15)

Even after fifty years of a steady rain of scientific and technical information on the damaging effects of smoking, the CEOs are still hedging in their communications about technical information. Bible still reserves for himself and his firm the possibility of holding "opinions" that need not agree with widely accepted knowledge. And in regard to public health, the last sentence, by deferring to but not necessarily agreeing with public health authorities, allows room for the industry not to resist putting statements as, "Smoking causes cancer" on cigarette packs without necessarily acknowledging the statements to be true or agreeing with them. Mr. Bible made a similar carefully worded statement about the addictive power of nicotine.

GRAPHICAL IMAGES

Words can be difficult enough to pin down regarding their meaning, interpretation, and ethical significance. Even more problematic than words but of equal importance, especially for technical communication, is graphical images. In this section we will see how the meaning of graphical images relating to smoking can be difficult to pin down, equally so for their ethical significance.

To see how difficult it can be to pin down the meaning of a graphical image, take for example a photographic image of the steel magnate Andrew Carnegie of the early twentieth century. It could have as many different meanings and interpretations as there are people viewing it. To some people it might mean the quintessential success story of the self-made man achieving everything he put his mind to. To others it might signify a story of discrimination, a white male achieving great success in a field and in a cultural context that denies the same opportunities for success to those who are not white or not male. To some it might signify the whole purpose of laissez-faire capitalism—great individual success that leads to great industrial advances and economic success for the entire country. To others, though, it might signify the exact opposite, the dangers of laissez-faire capitalism with its unfair and socially damaging maldistribution of opportunities and wealth.

These meanings could be those intended by the communicator, or those of the intended receivers of the communication, or those of unintended receivers. In addition, the meaning we might ascribe to an image can be strongly influenced by the context in which it appears. In the example of Carnegie, the same image might mean one thing at the stockholders' meeting of a steel corporation but something quite different in steelworkers' union hall.

The same is true for the ethical significance of images. Take for example the Joe Camel image that was one of the most popular yet notorious advertising ploys of R. J. Reynolds Tobacco Co. (RJR). We are all familiar with various forms of Joe Camel's happy face from billboards and magazines from the 1980s until the mid-1990s. To many people, looking at it casually, Joe's image seems fairly innocent. It projects a socially desirable image, what RJR calls a "smooth" character, who is self-confident, attractive, independent, a trendsetter. It also has the charming goofiness of a cartoon image, which perhaps reinforces our initial feeling that it is not meant to be taken seriously. Perhaps we might even be silly to take it seriously.

Joe's image looks different on closer ethical examination, however. Yes, it is initially charming and silly, but we should ask what interests are served in having us feel this way? Does RJR simply want to make us feel happy? Are they selflessly spending millions on billboard images just to brighten our day?

Actually, part of the aim of the image is its charm, silliness, and innocence, which so captivates our attention that it diverts us from the grim realities of the advertised product. We are less likely to read carefully the required fine-print notice warning us that all cigarettes, including Camels, cause terrible diseases and death. We are diverted from the cigarettes themselves, too, and all the misery that they have brought and will continue to bring to millions of people. We are distracted from thinking of the cigarette in Joe's mouth as a "coffin nail," as it was known even decades before any surgeon general's reports. Furthermore, Joe's blithe happiness encourages us to feel happy and discourages us from feeling worried, grim, or even skeptical. It also suggests that if we were to have any attitude other than Joe's blithe happiness, then we are only being prudish party poopers, too like a "grown-up."

We do not usually think of advertising images as technical documents, but that really is what this image is about, though in an inverse way. It is, in effect, an *antitechnical* document that exists only as a result of a prior context of serious, technical documentation that it is meant to counter. The massive body of technical information on the health effects of smoking, the addictiveness of nicotine, the protracted legal arguments about "causation," the surgeon general's legal requirement for warning labels, all need to be countered if tobacco companies such as RJR are to remain viable. The Joe Camel advertising image counters these serious technical concerns by being the opposite: antiserious and antitechnical. The character is carefree, self-indulgent, living for the moment rather than gravely concerned about addiction and cancer years later. No technical facts are presented, no numbers, no warnings, no concern about disease and death, no statistical probabilities, no parentlike admonitions against smoking. And it is an image of a character concerned above all with "image," with being perceived as cool, "smooth," attractive, and socially dominant.

It is an old rhetorical ploy. The charm of diversions is an old theme running throughout the history of rhetoric and ethics. The unethicalness of those communicators who tell others only what they want to hear or who urge them to be cheery and to do as they wish is an old story. It runs through practically every one of Plato's dialogues, going hand-in-hand with the popular but sadly mistaken mockery of Socrates' ethical seriousness. Plato, recall, also said that rhetoric and ethics are inseparable. Plato and classical Athens are not really far off the mark because this is basically what the potential case of Janet C. Mangini vs. R. J. Reynolds Tobacco Company in California, which was set to go to trial in late 1997, was all about.

Mangini claimed that the Joe Camel image was an integral part of a conscious and willful plan by RJR to entice children and adolescents to become addicted to smoking tobacco in violation of federal laws against tobacco advertising aimed at minors. RJR also deliberately deceived Congress about their advertising program, Mangini claimed. Before the case went to trial, however, RJR and Mangini settled out of court for $10 million and an agreement from RJR to cease Joe Camel advertisements but without any admission of guilt or wrongdoing.

R. J. Reynolds over the preceding four decades had been facing declining sales and the prospect of continuing decline. One of the chief reasons was that it was losing smokers and not gaining "replacement smokers." Long-time smokers were steadily dying off, either naturally or prematurely due to the diseases of smoking. Others just quit. These lost consumers were not being replaced because of the success of campaigns against smoking and of campaigns aimed at young people specifically.

RJR had conducted extensive research over the years on how smokers developed their smoking habits. They knew that 90 percent of smokers started to smoke by the time they turned 18, often years before that. They also knew that most smokers were very loyal to the first brand they had smoked and were very resistant to changing brands later in life. Therefore, in order to obtain a new band of smoking consumers who could be counted on to purchase RJR products for decades to come, it had to hook them early, that is, at or before age 18, with their product specifically. In a 1974 presentation to RJR's board of directors, for instance, vice-president for marketing, C. A. Tucker, explained, the "young adult market represent[s] tomorrow's cigarette business. As this 14–24 age group matures, they will account for a key share of the total cigarette volume—for at least the next 25 years" (Mangini at UCSF, 1–2). A senior RJR researcher had said earlier, "Realistically, if our company is to survive and prosper, over the long term we must get our share of the youth market" (2).

RJR in the 1970s in its French operations had revived the Joe Camel image, which it had used earlier in World War II advertising. The image was updated and made younger and cartoonlike to "youthen" the image of Camel cigarettes in order to appeal to a younger audience. In Canada RJR conducted research on advertising to potential smokers in the 15–17-year-old range. They learned that the Joe Camel image could appeal to an audience as young as 9 years old and entice them to begin smoking.

Part of the attractiveness of the Joe Camel image to youths is that it resonates with many of the attitudes that young people have—rebellion, defiance of authority, carefree behavior, absence of restraint and forethought, and a strong desire to have fun and avoid seriousness. Another important part of its attractiveness is simply that it is an image. Research had shown RJR that teens and younger audiences paid little attention to textual messages or factual information, preferring instead images and the emotions they evoked. To the youngest audience, of course, the cartoonlike appearance of Joe was naturally attractive and preferable to realistic likenesses.

Young people typically are impressionable, easily led, short sighted, and little concerned with critical thinking; and the younger they are, the more so they are. The industry deliberately aimed its advertising to the age group least likely to resist their messages and least likely to be concerned with their health whether immediately or in later years. These are only some of the reasons that advertising about smoking to minors is specifically prohibited by federal law.

It might have been argued, had the trial begun, that the tobacco industry was only vigorously protecting its interests and profitability, which on the face of it seems understandable enough. Certainly organizations are generally assumed to have a right to exist and to work to ensure their continued existence. There are ethical exceptions to this assumption, however. No one would argue that the Nazi regime was justified in pursuing its continued existence at any and all costs, for example. Similarly for an illegal organization such as a narcotics cartel. Though seeing their activities as aimed at self-preservation makes the activities seem understandable, it does not necessarily make them justifiable or ethical.

Steven Katz, Dale Sullivan, and Carolyn Miller among others explain in detail the potential ethical difficulties when we think that the pragmatic concerns of a technical enterprise, such as technical excellence, expediency, or utility, are in themselves *sufficient* grounds for justifying its actions. Katz's article about the Nazis, for instance, explains that Hitler undertook to make the values of technical excellence and technical expediency into a morality justifying all sorts of abominable political policies. What is excluded when one focuses solely on the technical means for achieving particular goals, however, is an ethical concern about whether those goals are good and desirable. Sullivan argues similarly that all too readily a concern for technical excellence can dominate all other considerations and all other ethical values. In the case of the tobacco industry, concern for their continued existence and profitability displaced an ethical concern about the goals of its advertising. Miller likewise argues that technology can become a form of consciousness informing our world-view, our sense of ourselves, and our sense of what is good and desirable. This dominating influence can, though, delude us from conventional ethical concerns. In the case of the tobacco industry, advertising can be seen as a technical activity, a collection of techniques that can be honed to be as effective as possible a means to achieve particular goals while avoiding any ethical concern about the goodness of those goals.

Another ethical lesson about the use of graphical images by the tobacco industry concerns the famous Marlboro Man. Richard Kluger in his Pulitzer Prize-winning history of the American cigarette war, *Ashes to Ashes*, uses the example of a Marlboro advertisement that appeared in the February 2, 1974, issue of *Newsweek*. It shows a pair of smoking horsemen against birch trees clothed in snow. The pure, clean, white snow and open countryside surrounding a healthy, rugged cowboy actively roaming the countryside presents an image that is the *exact opposite* of reality. The exterior images distract from interior effects. The reality is that smoking pollutes the lungs and leaves the body sick, weak, even cancerous, bringing with it debility and death. Life for a lung cancer victim is concerned entirely with the interior in several senses, rather than with external appearances and enjoyment of the outdoors. The beautiful winter scene also covers the entire face of the huge billboard, completely undercutting, overpowering, and trivializing the small, textual warning required by law. This is not just a diverting but a pointedly oppositional one. Recall that the sophists were renowned for deliberately opposing conventional thinking, and were criticized for making bad things appear to be good.

Sadly, but not surprisingly, that paragon of manly smoking indulgence himself, David McLean, also known as Marlboro Man, in 1995 succumbed to smoking-caused lung cancer and emphysema. His widow by 1998 was in the process of suing the Philip Morris Tobacco Co., together with several other major tobacco companies, for having caused his wrongful death and personal injury through fraud, deceit, negligence, misrepresentations to consumers, and breaches of warranty. Mrs. McLean contends that her husband was addicted to smoking and so could not stop even when he knew it was harming him. In addition, she contends, the tobacco companies withheld health hazard information from Mr. McLean as well as the public and misrepresented their product as not being hazardous to health. She also contends that Mr. McLean was forced to engage in harmful activity in the form of smoking up to five packs of Marlboros per day in order to get the ash to hang just the right way for attractive advertising or to get the smoke to curl just so.

Graphical images can work for the other side, too, fortunately. In the Mangini settlement, for instance, 90 percent of the $10 million given to the claimants is earmarked for anticigarette advertising directed to the same audience of youths. Happily, rather than using stern, adult-type warning messages, the producers are learning from the tobacco industry's experience by using formats that appeal strongly to the youthful audience, such as rappers and cartoons.

ETHICAL APPRAISAL

In this chapter rather than in the others, much of the ethical commentary and appraisal has been presented within the body of the chapter as various documents were discussed. This section draws these ongoing comments together and focuses on specific ethical perspectives.

Aristotle. There are several dimensions of an Aristotelian perspective relevant to the tobacco documents. Considering the pursuit of truth as a virtue, there is no doubt that smoking is hazardous in causing cancer and other diseases and disorders. Practically the only people in the United States who do not "believe" this are the handful of tobacco CEOs. Their deliberate avoidance and suppression of knowledge and their control of research indicates that the tobacco companies do not wish to find the truth except in an impossibly narrow, selective, and self-serving sense. There is also little evidence of openness to the truth as learned by others, and plenty of evidence showing active efforts to keep themselves in the dark. In some other companies producing other products, such behavior might be regrettable but go overlooked. With this particular product and its massive damaging effects on the public health, however, this attitude and behavior are clearly unethical.

Considering virtue in the sense of honesty, clear thinking, courage, or responsibility, the tobacco companies' behavior is also unethical. They lack honesty, fail to reason according to legitimate prevailing standards of inference, avoid making difficult decisions and facing the painful truth, and irresponsibly sacrifice the health and lives of millions for the sake of personal and corporate profit. The industry has tried to represent itself as virtuous in insisting on the highest, strictest standards for the demonstration of proof and for the inference of causality. These standards, however, are inappropriate and unrealistic. The vast majority of the disinterested scientific community, including all the governmental authorities charged with protecting the public interest, reject the industry's contorted reasoning. Furthermore, what the industry represents as a "debate" or "controversy" lacks honesty by grossly distorting the state of authoritative opinion in the field.

Considering the nature and aim of rhetoric, the "tobacco wars" as they have been called appear perhaps to demonstrate the Aristotelian notion of adversarial rhetorical debate between opposing parties. Recall that Aristotle explains that the chief value of rhetorical debate is that it forces a decision to be rendered between earnest, opposing parties. This conflict allows for the genuine truth of the matter to assert itself naturally, like cream rising to surface. Our judicial system in the United States is founded on this principle, as the defense and the prosecution take turns making their cases before a judge and jury. Actually, the issue is coming to just that as the supposed "debate" is moving at last into the formal legislative and judicial forums charged with negotiating such disputes. Though the industry's claim that debate is an essential element of scientific inquiry is true, over the decades practically the entire mass of reputable research evidence has weighed against them. In light of this disparity, their contention that smoking does not cause cancer amounts more to a sophistic contentiousness than Aristotelian debate seeking the truth. Indeed, Aristotle himself used an example from science to illustrate what matters are not suitable for rhetorical debate, that is, those in which the truth is apparent already.

Kant. From a Kantian perspective, you should act ethically in such a manner that you would wish your actions could become a universal principle applying similarly to everyone. From this perspective, the tobacco documents are clearly unethical. Repeatedly over the last four decades, a host of reputable, careful scientists, researchers, and governmental authorities have warned about the very real health hazard of smoking. These are people of unimpeachable motives, concerned for scientific truth as well as for the public health. They served the common good by revealing what might be inapparent to the public, that smoking causes diseases and that nicotine is addictive. It is fair to say that these careful researchers are treating the public as the researchers would want themselves to be treated. Time after time in the literature, researchers who themselves were smokers at the time their research began gave up smoking as their results became clear and undeniable. These researchers, in making their reports, only want the public to come to the same realization that they did.

The tobacco industry, on the other hand, has done the opposite. It has continually opposed and tried to undermine those who were working for the public good. They continually misrepresented information and the state of research generally; insisted on unrealistic and inappropriate standards for establishing proof and causality; and wilfully muddied the issue in the minds of the public. They deliberately tried to confuse the public through the endless repetition of empty claims. They have also knowingly withheld valuable information that could have benefited the public health. They have sought to confuse the public also by arguing in the manner of the sophists that a mountain of evidence is not a proof and that a cause is not a cause.

Their motives and the ends they sought were only the continued existence and profitability of their industry, regardless of the public good. It is clear that the industry did not treat the public or other researchers as it would want to be treated itself. No doubt the real, individual people within the artificial, corporate entity of the industry would not want themselves treated this way in any matters relating to their own health. They would want to know what is dangerous and addictive and would want any known sources of harm removed from their lives. Even more important when it comes to their own medical care or that of their families, they would want researchers, doctors, and the government to err on the side of caution and health.

Utilitarian. The utilitarian perspective weighs costs against benefits. In applying this perspective to our topic, the fundamental question is, To whom?

As far as the public is concerned, the costs, whether monetary, social, or emotional, vastly outweigh any benefits. Indeed, it is hard to conceive of any real benefit to tobacco smoking at all. Several state governments are in fact suing the tobacco companies precisely to recover the enormous monetary expenses they have incurred for treating diseases caused by tobacco use. It is true, though, that on some occasions the tobacco industry and its supporters have argued that tobacco use should not be restricted or banned because reduced tobacco sales would result in the negative social consequences of unemployed tobacco workers and reduced tax bases for the states involved. Few outsiders find much merit in such arguments.

To the tobacco industry, of course, the weighing of costs and benefits has always been the basis of its strategy. The industry simply wants to maximize its benefits while limiting its costs (and liabilities). Few people outside the industry, however, would accept this sort of analysis as being in any sense legitimately ethical.

Feminist Perspective and Ethics of Care.   Most feminist ethical thinkers are greatly concerned about impersonal corporations driven only by their own closed system of values and concerned only with its own self-interests. They would argue that the stonewalling and deliberate deceits of the industry show capitalism and free enterprise at its worst. Many feminist critics, as well as Marxist critics, are highly critical of the way U.S. culture grants legitimacy to impersonal entities. The resulting impersonal treatment of others, whether as the public or as workers, by corporations can only be damaging to them, they point out. Furthermore, the tobacco industry's reliance on arguments supposedly founded in scientific scrupulousness reveals perhaps one of the worst sides of science as an enterprise. It also shows how silent values in the background play a powerful role in how science is carried out. Quibbles about the validity of "the science" are really beside the point.

From an ethics of care perspective, the tobacco documents are appalling. They show a knowing, deliberate rejection of any responsibility for the care of those using their products. This has occurred not once but continuously over a period of decades as a conscious industry-wide policy. Of all the ethical perspectives we have considered, this perspective would emphasize the wilful irresponsibility of bringing years of debility and slow, misery-filled death to millions.

Topics for Papers and Discussion

1. You are a technical communicator and Web master for a major tobacco company. Your company has been ordered to provide research reports, internal correspondence, and any material relating to its knowledge of the health effects of smoking and the addictive power of nicotine. You have been given the task of coordinating a group of technical communicators to post these documents to a special Web site. Your supervisor has given you an enormous mountain of documents to which an endless stream of new documents is added each day.

You are told to index these documents using an arbitrary numbering system that gives no indication of the content, purpose, or significance of the document. This will make searching the documents very difficult and time consuming, of course. You begin to suspect that your company is dragging its feet in order to gain time to develop a new defense or damage control strategy while appearing to comply with the court order. You *are* making the material publicly available, after all, though only in a way that obscures its meaning and minimizes its accessibility and usefulness.

You have several good ideas for making the Web site more usable by government investigators, potential liability claimants, and the public. You feel it is your ethical responsibility as a technical communicator to do your job as well as possible and so to make your Web site as usable as possible for viewers, as recommended by the code of ethics of your professional organization. On the other hand, your professional organization's code of ethics also recommends that you comply with your employer's wishes as much as possible in performing your assigned tasks. What do you do? Why? What are the values and issues at stake for you?

2. You can research many interesting and timely topics that would show either the indefiniteness of supposedly objective facts or the conflicting usage of the same term and concept by different parties. Some of these topics are given in the following list, but you might prefer to research a topic from your own field of interest or expertise.

Write a report on your research, and make a brief oral presentation on it to the class.

a. The threshold for *safe* exposure of humans to radiation has shifted over the years. What is safe one year might not be safe the next. Research some of the changes in radiation safety criteria over the period 1950–1990 by the Atomic Energy Commission (AEC), now the Nuclear Regulatory Commission (NRC). Report on how the interpretation of objective information changes over time. Identify the interests at work that seem to have motivated these changes. What are some of the different meanings and definitions of "safe" that you encountered? What were some of the issues raised in arriving at various definitions? As an added dimension, consider that some scientists say there is *no* safe level of exposure to radiation of any kind. What would be some of the industrial and economic effects of making this statement the basis of government regulations?

b. The health effects of leakage from silicone breast implants is another situation in which the meaning of supposedly "hard" technical and scientific information, including statistics and probabilities, has been hotly contested. Dow Corning Corporation has been ordered by the court to pay billions in settlement of health claims against it for having manufactured these implants (and, as a result, has been reduced to bankruptcy). The general opinion of the scientific community, especially of those most knowledgeable in this area, however, is that the breast implants did not *cause* the health effects that have been attributed to them. Serious and tragic though the health problems suffered by the claimants are, they feel that the health effects had causes other than the implants.

Research several articles, including editorials, in professional journals on this topic. Try to find differing positions showing how experts disagree in many ways on the "facts" of the matter. This topic is admittedly a sensitive one for many people. This would be a good opportunity to reflect on your own earnest reaction and those of others, considering the values that are operating in their responses. This will allow the discussion to focus on values per se, with the scientific and technical statements serving as a context for the discussion. This literature is highly specialized, and the sheer volume of it is daunting, so this topic would be best pursued by those with a background in a biomedical field or in statistics. A small group could divide the investigation into several sections.

c. The safe disposal of radioactive waste, especially waste of moderate and low intensity, is another hotly contested topic. The amounts are ever increasing, and the materials will remain hazardous for centuries, even millenia. Two proposals have been considered recently by the federal government, one involving forceful implantation into deep-ocean sediments, another into shafts drilled into a granite mountain range in the U.S. Southwest.

Research one of these plans in detail as a term paper, examining several different viewpoints. You will find that there is a good deal of earnest disagreement on the key technical "facts." The supposed stability of deep-ocean sediments or of geological formations is a matter of interpretation, probabilities, and speculation rather than absolute certainty. Summarize the positions and the evidence in support of the different sides, then describe the values or interests that seem to be driving the different representations and interpretations of technical information.

3. The movie *Silkwood*, produced several years ago, traces the discovery by an employee of a nuclear materials manufacturer of serious health hazards and the misreporting of required information. Research the actual events and the case on which the film is based, including specific technical documents (or the absence thereof). Describe in your report how this case demonstrates the unethical control of technical information and the denial of knowledge by a company. Note that a feminist perspective would highlight the hierarchical organization of the corporation, power, even violence as a means to control the dissemination of technical information.

4. The strict meaning and usage of the term "cause" continues to be the focus of opposition by the tobacco industry to efforts to curtail tobacco use. In 1991 the *Journal of the American Medical Association* (Vol. 266, No. 22), the best known and one of the most highly regarded journals of medical science, published a landmark series of articles on tobacco advertising, especially about Joe Camel, and its effects on the smoking practices of young people. These articles, as well as later ones in *JAMA* and elsewhere, provided the technical and scientific groundwork for efforts to further restrict tobacco advertising affecting young people (such as the Mangini case).

You might find it useful to research the efforts of the Federal Trade Commission (FTC) to eliminate cigarette advertising that is attractive to a youthful audience, in particular Joe Camel advertising. The FTC contends that Joe Camel advertising violates federal laws against advertising cigarettes to minors. Causation is important in this matter because of claims that advertising causes smoking behavior and while others claim that it does not. As one of the FTC commissioners explains, the crucial question is whether there is "a link between the Joe Camel advertising campaign and increased smoking among children, not whether smoking has an effect on children or whether the health of children is important." The FTC offers news and information on topics such as this at its Web site at http://www.ftc.gov. Copies of many actual FTC documents are available on-line or can be ordered from the Web site.

Trace the meaning and usage of "cause" in several of these scientific and technical articles. Describe how the authors rhetorically deal with possible quibbles about strict causation from critics. This reveals the authors' awareness and accommodation of their entire audience, both sympathetic scientists and doctors and the unsympathetic tobacco industry opponents. Describe how the authors cultivated their ethos (credibility) and carefully shaped their logos (reasoned argument), two elements of traditional rhetoric. What are the values, whether explicit or implicit, at work in these documents?

5. "A Frank Statement to Cigarette Smokers," published by the new Tobacco Industry Research Committee, appeared prominently in all major U.S. newspapers on January 4, 1954. Even though the statement was made almost half a century ago, it is one of the most frequently cited documents of the tobacco industry in suits against them in recent years, which have focused not only on product liability but also on outright fraud, in addition to other charges. As mentioned earlier, there are three reasons for its enduring interest. First, it sows the seed of misrepresentation and confusion on a crucial technical matter, namely whether causation has been scientifically proven. Second, it explicitly and prominently acknowledges not only that the industry has a basic responsibility toward the public, but that this responsibility is more important than all other business considerations. Third, it defines the purpose and composition of the Tobacco Industry Research Committee, which, it was claimed, was to assist in research into the health effects of smoking. Litigation and access to secret documents in the 1990s shows all these basic contentions to be false and inconsistent with the state of knowledge within the industry at that time.

Obtain a copy of "A Frank Statement" from newspapers of the time, from the Web sites of various lawsuits against the industry by state and federal governments, or from the Web sites or literature of various anti-smoking organizations (see suggested sites at the end of this chapter). Analyze the language and argument of it to show its intended effects. Pay particular attention to qualifiers or the lack therefore, which can leave misleading impressions. It would be helpful to research some of the history of the tobacco industry to control, contain, or deflect increasing concern about scientific and medical professionals about smoking. This will help you to better understand the context and grasp the truth-value claims made in the statement. Write a report on your findings and impressions.

As an advanced research project, obtain transcripts of the famous trial in 1995-1998 of the State of Minnesota and Blue Cross and Blue Shield of Minnesota versus various tobacco companies, in Minnesota District Court, Ramsey County. The opening statements for the plaintiffs and the defendants on January 26 and 27, 1998 respectively are particularly interesting because of the attention paid to "A Frank Statement," even forty-four years after it was published. These two statements show the various ways the language and meaning of the statement can be construed, the complex historical context of its claims, and the complicated later development of the issues identified in it. The transcript of April 15, 1998 is also interesting as B. Scott Appleton, director of scientific and regulatory affairs for Brown and Williamson testified about the relation of "A Frank Statement" to other, contradictory technical documents possessed by the industry. The questioning of Appleton highlights as well the crucial role of ethics in judging the validity of "A Frank Statement" and the ultimately personal nature of ethical judgments; ethics is not just for professional ethicists and philosophers.

Summarize the opening statements of both sides on "A Frank Statement," then present what would have been your judgment about the ethical worth of the statement had you been a juror hearing the trial. They can be found at http://www.tobacco.org/Documents.

REFERENCES

"A Frank Statement to Cigarette Smokers." *New York Times* January 4, 1954, 15.

Glantz, Stanton A., John Slade, Lisa Bero, Peter Hanauer, and Deborah Barnes. *The Cigarette Papers.* Berkeley: University of California Press, 1996.

Hanauer, Peter; John Slade, Deborah E. Barnes, Lisa Bero, and Stanton A. Glantz. "Lawyer Control of Internal Scientific Research to Protect Against Products Liability Lawsuits." *JAMA: The Journal of the American Medical Association* Vol. 27/3, July 19, 1995, 252–66.

Katz, Steven B. "The Ethic of Expediency: Classical Rhetoric, Technology, and the Holocaust." *College English* 54/3 (1992): 255–75.

Kluger, Richard. *Ashes to Ashes: America's Hundred-Year Cigarette War, the Public Health, and the Unabashed Triumph of Philip Morris.* New York: Knopf, 1996.

Koop, C. Everett, David C. Kessler, and Geroge D. Lundberg. "Reinventing American Tobacco Policy: Sounding the Medical Community's Voice." *Journal of the American Medical Association* 279/7 (February 18, 1998): 550–52.

Miller, Carolyn. "Technology as a Form of Consciousness: A Study of Contemporary Ethos." *Central States Speech Journal* 29 (Winter 1978): 228–36.

Old Gold Advertisement. *Life* 36/1 (January 4, 1954): 50–51.

Peto, Richard, Alan D. Lopez, Jillian Boreham, Michael Thun, and Clark Heath Jr. *Mortality from Smoking in Developed Countries, 1950–2000: Indirect Estimates from National Vital Statistics.* Oxford, Oxford University Press: 1994.

Sullivan, Dale. "Political-Ethical Implications of Defining Technical Communication as a Practice." *JAC: Journal of Advanced Composition* 10 (1990): 375–86.

WEB SITES

ASH: Action on Smoking and Health, a national legal-action antismoking organization: www.ash.org

"Inside the Tobacco Deal" accessible through the site of the PBS program *Frontline:* www.pbs.org/wgbh/pges/frontline/shows/ and through their fictional docudrama called "The Cigarette Papers": www.pbs.org/wgbh/pages/frontline/smoke/webumentary

"Tobacco Control Archives" accessible through Galen II of the medical library of the University of California, San Francisco: www.library.ucsf.edu/tobacco

Web Site of the House Commerce Committee, which houses the documents released under the Minnesota case against the tobacco industry (an almost unusable site because of uninformative indexing system). Fortunately, the PBS webumentary refers to specific documents and identifying numbers. The term "master" in this context refers to the judicial title of the person who reviewed documents that the industry wished barred from public disclosure. 143.231.226.4/tobaccodocs/master.htm

■ ■ ■ ■ ■

CHAPTER 7

STAR WARS
HOPE VS. REALITY

In this chapter we examine the ethical issues in technical communications involving claims about the software that was to be used in the Star Wars nuclear missile defense system. Though the formal name for this program is the Strategic Defense Initiative or SDI, it is commonly known as Star Wars. The crucial software would link all the other elements—such as lasers, radars, and missile interceptors—together and make them into a working system. It would thus conduct the actual "battle management." This vital element was referred to by one SDI director as "the long pole" that would hold up the entire tent of the SDI program. Though the SDI program itself has now been terminated, the research and development of various parts of the system still continue under many different, less ambitious programs. The testing of various shorter-range missiles intended to intercept incoming missiles, for instance, continues in the Theater High Altitude Area Defense system. Strategic ballistic missile defense continues under the Ballistic Missile Defense program. Much of the SDI computing activities have been redirected into the Accelerated Strategic Computing Initiative for nuclear "stockpile stewardship." Renewed interest and funding commitments to develop a limited defense against only a few missiles has followed the recent Rumsfeld Commission finding of a real threat from "rogue states" in the near future. Nevertheless, the Star Wars program has ended, and so has the heated debate about its technical feasibility.

Because the SDI program was never even partially deployed, much of the technical communications about the software planned for SDI have to do with the prospective future rather than the real, immediate present. These technical communications—and the technical communicators behind them—therefore had to speculate about what would develop in their field rather than simply presenting objective information about existing devices. This sort of technical communication is not really unusual. Nearly all practical communications are purposeful, aimed toward achieving future goals. For that reason they cannot help but support or express certain values, which in turn influence the shape of the information portrayed in them. In the case of Star Wars, however, this speculation seems to have gone too far. Though the program clearly struck a nerve in offering to the public a hope against the threat of nuclear annihilation, this hope seems to most observers to have been unfounded on realistic technical possibilities, even going so far as to speculate about technologies that had not even been discovered yet. In addition, some of these claims strongly suggested a present technical reality that simply did not exist.

Such claims are not necessarily unethical, however unrealistic they might be. On the other hand, it might have been more ethically responsible and in keeping with standard governmental practice to have consulted more thoroughly with recognized technical and scientific experts on the feasibility of claims made for Star Wars. Furthermore, claims about the feasibility and effectiveness of Star Wars were couched in language that freely intermixed past and present technical reality with future goals and wishful expectations. Such statements confused hope with reality.

We will see how claims made about the ability of the proposed software to operate the Star Wars system were exaggerated, selectively represented, or misrepresented. We will see too how vague and misleading such technical communications can become when they deal with hopes and speculations rather than with realistic expectations. We will also see how several scientists and technical experts finally voiced their ethical opposition to these distorted technical communications and so played an active role in the eventual demise of the program.

In the process, we will come to see how interests and attitudes—specifically national security and technological optimism—can be carried too far. National security is naturally of paramount importance to both national leaders and the general public, as it should be. Technological optimism has its place, too, in fostering new technical achievements. But though rhetorical appeals grounded in such powerful concerns can motivate us to prudent action, they can also be applied to ethically questionable purposes.

The mission proposed for SDI was incredibly complex. This complexity lies at the heart of the software issue and so will be explained first before we examine specific technical documents. The mission assigned to SDI changed substantially over time, too. The speculative nature of the technology and its shifting mission are part of the reason it is difficult to say anything definitely about the program. This indefiniteness can be both an advantage and a disadvantage. It allows proponents to have it both ways, in effect. They can contend that their statements are realistic because they are built on past and present technical realities. On the other hand, they can also contend that future expectations for the program are beyond any criticism grounded merely in present realities.

First we will review the overall context in which Star Wars came to life in order to grasp the compelling force of the problems it undertook to solve. This will be followed by an outline of how it was proposed to work, showing the extreme complexity and speculativeness of the program, which gets to the root concerns of those opposed to the program. The next section will summarize reports on SDI by the congressional Office of Technology Assessment and present one example of congressional hearings on SDI, including testimony by

Senator John Glenn. Two subsections will follow this, one of statements supporting the program, the other of statements opposing the program. The most important of the latter is the formal objection to the program by David L. Parnas, the founder of the field of software engineering. The Star Wars Boycott Pledge will be discussed too. This pledge was signed by thousands of the country's leading scientists and engineers, those most familiar with the technical feasibility of the program.

In this Star Wars case, in contrast to the Nazi and tobacco industry cases, the documents and their drafters that we will review were not necessarily unethical because debates on important public policy issues commonly are two-sided, both sides having legitimate concerns driving their statements. Our democratic process assumes this. On top of that, practical communications aim toward achieving specific purposes that must represent interests and values. Therefore practical communications naturally support and express values and aim toward goals defined by these values. Values also shape the portrayal of technical information and the nature of claims stemming from technical information. For this reason we should not be surprised that values played prominent roles in the Star Wars documentation. What is surprising, however, is the depth of this influence, which in some cases seems to have been excessive and unwarranted though not necessarily unethical.

CONTEXT

Everyone instinctively wants to be safe from harm. Who, then, would not want to be safe from the most destructive devices ever created, nuclear weapons? Safety and protection were the focus of the rhetorical appeal that President Ronald Reagan presented to the American people on March 23, 1983. In this famous speech President Reagan presented his plan for the Strategic Defense Initiative (SDI), which came to be known as "Star Wars," after the movie of the same name.

The motivation to try to develop a defense against nuclear weapons is obvious and deeply felt. As one senator noted in his testimony to Congress about SDI, every U.S. president since the beginning of the nuclear stalemate between the United States and the Soviet Union has faced a simple but profound realization. He may wake up in the middle of the night cold with fear that the Soviet Union has thousands of incredibly destructive weapons trained on the United States. He wonders what can be done to stop an attack should one be launched and realizes that the answer is simple: Nothing. He immediately decides that this is an intolerable situation and takes action to try to change it. So the motivation is obvious. But the technical feasibility is not obvious nor is the likelihood that somehow totally new technologies can be created on demand.

The statement that created SDI sketched a plan of enormous scope, but as a sketch of a technical program it suffered an important difficulty: It was almost entirely a statement about goals. It was not clear exactly who or what would be covered by the plan, or when, or even how. The reason for this indefiniteness was that there simply was no means then (or now) available to achieve this goal. Because of this, though the plan calls for research and development, its language strongly suggests, if not states, that the goal is achievable and soon due to new technological developments. Thus what seems highly appealing at an emotional level and what seems plausible at a stated technical level turned out to be technically impossible. And so tens of billions of dollars were spent on a program that was projected eventually to cost trillions, before it was terminated.

The documents we will examine show the vitally important role of values in the representation and interpretation of technical information. Seemingly definite, objective topics such as computer software programs and the performance we can expect of them often carry with them a heavy burden of judgment, both technical and ethical. The congressional Office of Technology Assessment, for instance, went to great lengths to try to pin down the precise meaning of the term "software reliability" in the context of SDI. They concluded that the term has many facets, both definite and indefinite, but that it was essentially impossible to determine precisely and unequivocally the reliability of the software planned for SDI.

We should understand that our appraisal of these documents is somewhat problematic. Our discussion is not about whether nuclear weapons are terrifying or whether the world would be a better place without them. The question is only whether appeals to national security should be allowed to exaggerate, distort, or misrepresent realistic technical matters. We should also recognize that all the documents we will examine are unclassified. Though they might not give us the whole picture, they are generally accepted as outlining the terrain. We will take these documents at face value because that is all we have to work with.

OVERVIEW OF SDI

In this section we will see how extremely complicated the entire system would have been. By understanding the extreme complexity of this system, we can better appreciate the central importance of the software that would operate it. We will also see how the SDI program changed over time in its elements and mission. These changes reflect the indefiniteness that surrounded the program because it was more a hoped-for goal than a realistic technology.

President Reagan dramatically announced to the American people on March 23, 1983, on live television that he was initiating a program to defend the United States against nuclear attack. This program would remove forever the threat of nuclear annihilation that had hung over the country—and the whole world—since the beginning of the nuclear era. It was a startling announcement because of its vast scope, because it was unanticipated by the public, and because of the astronomical expense of the proposed system. Many critics consider it one of the most important military statements ever made by a U.S. president to the public.

In this speech President Reagan explicitly challenged the American scientific and technical communities to develop ways to intercept and nullify any nuclear weapons aimed at this country. The proposal stems from what is known as "technological optimism." This is the assumption that any problem can be solved by technological means and that this technology can be devised if only sufficient will, money, and talent are applied to the problem. Though the specific means to accomplish the SDI goals were for the most part unknown, President Reagan counted on the ingenuity that the United States had earlier demonstrated in other monumental scientific and technical undertakings such as the Manhattan project that developed the atomic bomb and the Apollo project that brought humankind to walk the face of the moon.

The goal was clear: "to intercept and destroy strategic ballistic missiles before they reached our own soil or that of our allies" and to render these nuclear weapons "impotent and obsolete." How was this to be accomplished? The speech is not entirely clear, offering not so much a plan but a tantalizing goal in the form of a "vision of the future." Though the details are not spelled out, the goal is presented as technically feasible. This feasibility is supported by an attitude of technological optimism that runs through the speech. If new technologies are needed, given the "can do" spirit of American ingenuity, then they will be created:

> The solution is within our grasp. . . . Let us turn to the very strengths in technology that spawned our great industrial base. . . . I know this is a formidable technical task, one that may not be accomplished before the end of this century. . . . Current technology has attained a level of sophistication where it's reasonable for us to begin this effort. . . . America does possess—now—the technologies to attain very significant improvements in the effectiveness of our conventional, nonnuclear forces. Proceeding boldly with these new technologies, we can significantly reduce any incentive that the Soviet Union may have to threaten attack against the United States or its allies. . . . I call upon the scientific community in our country, those who gave us nuclear weapons, to turn their great talents now to the cause of mankind and world peace, to give us the means of rendering these nuclear weapons impotent and obsolete. . . . I believe we can do it.

The first step in reaching this goal, the president explains, is to begin research and development on these technologies in the SDI program: "I am directing a comprehensive and intensive effort to define a long-term research and development program to begin to achieve our ultimate goal of eliminating the threat posed by strategic nuclear missiles."

The speech calls for a program to make nuclear weapons "impotent and obsolete." This was understood by practically everyone as calling for the total protection of everyone in the United States against missile attack. As time went on, however, reality clashed with hopes and claims. Complete, "perfect" protection stopping 100 percent of all enemy missiles was later reduced to "near-perfect" protection, then "less-than-perfect" protection involving maybe 400 megatons "leaking" through SDI protection. In addition, rather than protecting all of the United States including Alaska and Hawaii and its allies as the speech indicated, the program came to be limited to only the continental United States (excluding Alaska and Hawaii) and still later only a few major cities and military sites. By then the scope was limited to an attack by only a few missiles from "rogue states" or from unauthorized launches of Soviet missiles by deranged military officers.

The speech is noteworthy for its ethical and moral components, too. Coupled with the goals of the program and its technical optimism are clear statements of the ethical values driving this technology:

> I've become convinced that the human spirit must be capable of rising above dealing with other nations and human beings by threatening their existence. . . . Wouldn't it be better to save lives than to avenge them? . . . Are we not capable of demonstrating our peaceful intentions by applying all our abilities and our ingenuity to achieving a truly lasting stability? I think we are. Indeed, we must.

Thus the SDI program was linked to the values motivating it and defining its goal. This value dimension gives the speech its ethical thrust. The possible relief from fear of the threat of nuclear annihilation gives the speech its emotional thrust. Clearly the goals are desirable and worthy, but were they technically feasible?

After we review the complexity of the overall system, we will examine statements by representatives on both sides trying to answer this question. Specifically we will examine statements about the feasibility of a trustworthy, realistic software program to act as the "glue" holding all the parts together in a working system. We will see that the weight of expert opinion in the years following this speech supports the judgment that the answer is, No.

The point of our examination is not that any of the statements in support of SDI are necessarily unethical, however. A difference of opinion does not assume one side is unethical, after all. In addition, our democratic system of government assumes that there will always be open debates between opposing sides on almost any issue, sides that are sincerely, reasonably convinced of the goodness and rightness of their positions. The purpose of these examples is only to show the powerful influence of values in shaping public discourse and technical claims about a highly technical topic.

A Complex System

The basic structure of the program was first defined by the report of key scientific and technical advisers, known as the Fletcher Report. Its eight key areas were later reorganized into five when the Strategic Defense Initiative Organization (SDIO) was formed in 1986. These five areas are surveillance, acquisition, tracking, and kill assessment; directed energy weapons; kinetic energy weapons; survivability, lethality, and key technologies; and systems concepts and battle management. Systems and battle management includes all the means by which elements such as kinetic weapons and directed-energy weapons would be integrated into a system and operate as a unit. Its battle management function consists of command, control, and communications (BMC3) and its required computer hardware and software. This software is the crucial link that would make the whole system work, or not. Lieutenant General Abrahamson, the first director of SDIO, used the metaphor of a military tent to explain the crucial importance of this software, which he called "the long pole" holding up the whole SDI tent. Claims about this software are the focus of our ethical examination.

It is important for us to understand the great complexity of the SDI system because this complexity dictates the demands of the software that would integrate the elements of the system. (A good guidebook on the complexity of SDI and the debates about it for a nontechnical audience is *Claiming the Heavens*, by Philip Boffey et al.)

SDI involves the interception and elimination of incoming intercontinental ballistic missiles. The flight of these missiles has four phases: boost, postboost, midcourse, and terminal. Interception can take place at any of these four phases. Plans for interception involved redundancy by which interception would be attempted at several phases or tiers in order to ensure success.

A typical flight begins with a booster burn of two or three minutes. This is followed by a postboost phase of about five minutes in which one or more warheads are aimed toward their targets. Decoys and other deception devices are then deployed. During the midcourse phase, these warheads coast for about twenty-two minutes. By then the warheads are traveling at over 10,000 miles per hour in airless space at an altitude of several hundred miles. About thirty minutes after launch, the warheads reenter the earth's atmosphere in the terminal phase, which lasts about one minute.

An incoming missile was to be detected at the moment of launch by orbiting surveillance satellites looking for the large plume of hot exhaust gases. (Some of these sensors have already been developed and are in orbit now. The tragic explosion of TWA Flight 800 over Long Island Sound on July 17, 1996, was detected by one of these satellites.) Identification and location would be refined by additional satellites alerted by the surveillance satellites. A system of space-based lasers in permanent orbit would then blast the booster with a beam of extremely intense light. Later versions of SDI included a "pop-up" X-ray laser launched from submarines in constant patrol around enemy territory. These lasers would be powered by nuclear explosions (which would also destroy the laser device itself of course). Due to the short time frame, interception would have to be entirely computer automated, without human control over the process.

During the postboost phase, warheads are dispensed from a "bus" carrying up to ten warheads and directed toward their targets. The bus and its warhead were to be detected by other surveillance satellites, a more difficult task because there would be no large exhaust plume showing. Orbiting lasers would then be directed to destroy the devices. These would be supplemented with even more powerful ground-based lasers that would bounce their beams off mirrors in space toward their targets.

During the postboost and midcourse phases, an orbiting swarm of one thousand or more space interceptors would attack the warheads. A later form had fewer satellites, called Brilliant Pebbles, that would be directed by fifty or so Brilliant Eyes satellites. (These devices have actually been developed and nearing deployment under the BMD program as of 1999. A decision will be made in June 2000 whether to adopt the latest national missile defense, which could cost $120 billion, Paul Mann reports.)

During the midcourse or coasting phase, the separate warheads would be very difficult to track because they are smaller targets and are dispersed on diverging trajectories. These would be detected by ground-based radars or by satellites. Interception could be undertaken by several means including orbiting or ground-based lasers, direct impact interceptors launched from satellites or from airplanes, or magnetic rail guns firing small guided interceptors. This defensive task would be complicated by various deception devices that the enemy might use including swarms of decoys, chaff to confuse radar, or lasers or radio transmitters to blind or confuse our sensors.

In the last or terminal phase, decoys and chaff would be nullified as atmospheric drag separated them from the heavier warheads. During this last minute or so, the warheads would be tracked by ground-based, airborne, and space-based sensors. Interception would be accomplished by high-speed ground-based missiles and small guided interceptors launched from rail guns. This interception task would be greatly complicated by complex, powerful electromagnetic disturbances to the atmosphere following nuclear detonations during an all-out attack. These detonations could come from our own defensive weapons, or from the enemy by earlier warhead delivery to the target, or from intentional early detonations meant to confuse interception.

All these detection, tracking, guidance, and interception devices would have to be precisely coordinated by various computer systems, interconnected by reliable communication links also under computer control, and operated under a computerized battle management system for command, control, and communication (BMC3). It would have at most only thirty minutes to complete its tasks successfully in the flight of each missile. Under the initial SDI program, there would be thousands of missiles, many times that number of warheads (most missiles carrying several warheads), and many times that number of decoys in a full-scale attack (see Figure 7.1).

The Battle Management System Communicates With All Systems Of The Strategic Defense To Acquire, Track And Counter Ballistic Missiles

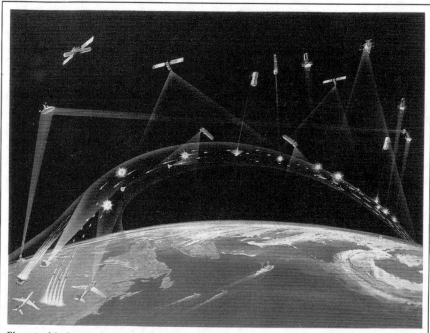

Elements of the Strategic Defense System are shown acquiring, tracking or countering ballistic missiles and their warheads in this artist conception. Attacking missiles launched at the right are detected and tracked by space-based sensors. They are countered by space-based interceptors and by ground- and space-based directed energy weapons from the boost to midcourse phase of flight. The warheads are continually tracked by space- and ground-based sensors through the midcourse phase. From the late midcourse to the terminal phase of flight, ground-launched interceptors begin countering the warheads.

FIGURE 7.1 Complexity of Star Wars This illustration shows the great complexity of the proposed Star Wars Defense System in an imagined mass attack. In order to prevent any warheads from reaching their target, all elements would have to be coordinated tightly and work flawlessly the very first time it is ever used. Only some of the planned devices are shown here. Not depicted, for example, is the controversial "pop-up" X-ray laser defensive weapon powered by a nuclear detonation, to be launched at a moment's notice from submarines constantly patrolling near the Soviet Union.

From Department of Defense publication *Strategic Defense Initiative: Progress and Promise*, p. 20.

Complicating its tasks would be various tactics the enemy might use such as intentional nuclear detonations in space, in the atmosphere, or on the ground to confuse sensors, disrupt communications, and "fry" computer chips. Sabotage could be directed at any or all of the links in the chain of BMC3. The mass launching of all enemy missiles at once might attempt to overwhelm the system. Alternatively, many successive launches over a long period might exhaust the interception systems, allowing later missiles to reach their targets unimpeded. Any and all elements of the system would also be under direct attack in many possible ways.

This computer system would have to operate in an extremely difficult environment that has no realistic parallel. For years it would have to orbit in the vacuum of space while subjected to extremes of heat and cold and radiation much greater than on earth. (Recall that mainframe computers on earth are housed in "clean rooms" under ideal circumstances of heat, humidity, and dust control.) For that reason, most critics agree that we cannot use existing computer systems as analogies for how the SDI system would function. The Federal Aviation Agency's system for tracking and controlling civilian air traffic, for example, operates in a very complex environment in which the target is not trying to avoid detection, with fairly predictable numbers and kinds of targets and in a peaceful situation with dependable power and communications and without nuclear detonations or deliberate sabotage.

The only remotely comparable analogies are some military systems such as Aegis or the World Wide Military Command Control System (WWMCCS). But these have poor records even under limited demands, much less under all-out nuclear attack. The U.S. Navy's sophisticated and expensive Aegis battle-management system has been touted by Pentagon officials as "Star Wars at Sea," yet confusion over identifying a single target near a regular commercial airway led it to shoot down an unarmed airliner, killing everyone aboard ("Star Wars at Sea"). In another notable instance, a new guided-missile cruiser in the U.S. Navy's "Smart Ship" program suffered a software glitch that shut down not only all its weapons systems but also its power, leaving it dead in the water for two hours ("Rough Sailing for Smart Ships"). Though recent reports suggest the problem occurred when changes were made to the operation of the software and do not reflect software problems per se, this actually agrees with concerns of SDI critics. They contend that SDI software would have to be changed repeatedly as weapons and policies changed, a process notorious for generating great problems. Clearly SDI poses an enormously complicated task for software engineering, a task that must be performed nearly perfectly under the most difficult of conditions in its first and only real use.

CONGRESSIONAL OFFICE OF TECHNOLOGY ASSESSMENT

A good place to start exploring the issue of SDI software claims is some of the official governmental statements on the matter. We will see that substantial difficulties were anticipated from the earliest days of SDI. These statements conflict with the highly optimistic statements by proponents of the program.

During the 1980s while SDI was a viable program, the official federal government organization for advising Congress on the latest and prospective technology was the Office of Technology Assessment (OTA), which no longer exists.

Just a few months after the president originated the SDI program, OTA issued a background paper on one element of the overall program, titled "Directed Energy Missile Defense in Space." These "directed energy" devices include X-ray lasers in space, chemical lasers on the ground bouncing off large mirrors orbiting in space, chemical lasers in continuous orbit in space, and neutral particle beams. The report also considers how these devices would be operationally utilized, in its sections on "command and control" and "battle management."

The tone is fairly optimistic but with significant concern:

> Command and control for BMD [ballistic missile defense] does introduce two interesting issues to which technology cannot provide an answer. The first is the impossibility of testing the whole defense system from end to end in a realistic wartime setting. . . . The BMD system would have to work near perfectly the very first time it was used. The second issue is the likely need for the defense to activate itself autonomously, since there would be no more than a minute for human decision. (41)

The report presents four "misapprehensions" regarding the stated goals of the president's plan:

1. Individual, separate devices such as lasers are not the same as the total system in which they would be used together, which would be extremely complex.
2. SDI is unlike any prior technical program such as the Manhattan project.
3. Hopes for entirely new technologies cannot be realistic. "Such breakthroughs are not impossible, but their mere possibility does not help in judging the prospects for the perfect defense" (67).
4. Accurate predictions cannot be made about the performance of this complex system. There is, and can be, no realistic test of the system beforehand; all possible outcomes cannot be anticipated.

All of these points are related to the software issue, we should note.

In September 1985 OTA issued another report on SDI. The executive summary (spanning 34 pages) states:

> Massive improvements in computer speed, reliability, durability in a hostile environment, and software capabilities would be required. Current research gives cause for some optimism about meeting the hardware requirements, though most analysts agree that generating the necessary software would be a monumental task (25).

The body of the report is even stronger, however:

> Specifying, generating, testing, and maintaining the software for a battle management system will be a task that far exceeds in complexity and difficulty any that has yet been accomplished in the production of civil or military software systems. . . .

> The problem of realistically testing an entire [battle management] system, end-to-end, has no complete technical solution. . . . There will be no way, short of conducting a war, to test a fully deployed BMD system (190).

A later section reiterates that the ten million lines of code of its program would have to function reliably and perfectly the first time it was used. The universal military experience, however, has been that large computer systems take years and a good deal of imperfect experience to debug them. The SDI system would also be used for the very first time under the worst possible battlefield conditions—all-out nuclear war—with sabotage, power outages, decoys, shock, confusion, and electromagnetic disturbance. Furthermore, these conditions have never been experienced before, so we have no real basis for effectively accommodating these conditions.

In May 1988 OTA reported on several aspects of SDI including the software requirements specifically. OTA again cited software as a key obstacle. By then, the hardware devices for intercepting missiles had become proportionally less important while concerns about the computer management of the system had become more important, covering fully half of this report.

The chapter on software draws these three conclusions from a total of eight:

1. The nature of software and our experience with large, complex software systems, including weapons systems, together indicate that there would always be irresolvable questions about how dependable the BMD software was, and also about the confidence to be placed in dependability estimates.
2. No matter how much peacetime testing were done, there would be no guarantee that the system would not fail catastrophically during battle as a result of a software error. Furthermore, experience with large, complex software systems that have unique requirements and use technology untested in battle, such as a BMD system, indicates that there is a significant probability that a catastrophic failure caused by a software error would occur in the system's first battle.
3. No adequate models exist for the development, production, test, and maintenance of software for full-scale BMD systems (249–50).

These reports show that from the earliest days of the program, serious concerns were voiced by technically knowledgeable government authorities about the feasibility of SDI and its software, though proponents were arguing that the program for a grand shield of protection was feasible.

CONGRESSIONAL HEARING

Of the many congressional hearings on SDI, we will sample only one in order to grasp the tenor of the debate. Keep in mind that, by design, the U.S. form of government deliberately sets the legislative branch, Congress, against the executive branch, embodied in the president, in a system of checks and balances. We therefore naturally expect that Congress would challenge the president's new initiative. Nevertheless, the debate over SDI was particularly heated and was particularly concerned about the possibly unethical misrepresentation of technical feasibility.

Our example is the Senate hearing before the Committee on Foreign Relations on April 25, 1984. This committee was interested in SDI because it would totally reshape our military posture toward the rest of the world, enemies and allies alike. It sought answers to two questions from each person testifying. First, does that person share the president's view that Star Wars would render nuclear weapons "impotent and obsolete"? Second, does that person share the view of Secretary of Defense Casper Weinberger that Star Wars would be a "reliable and total" system of protection against nuclear weapons?

First to testify was George A. Keyworth, the president's science adviser. Even then, just one year after the initiative was begun, its visionary aim had changed from total protection against nuclear attack to this: " 'We see the investigation of strategic defense options as an absolutely vital catalyst to real arms control' " (8). Its purpose now was not defense per se but leverage in negotiations for arms control and reduction agreements.

The representation of technological realities and possibilities was paramount in Keyworth's testimony. He cited and affirmed the judgment of the Fletcher panel's conclusion: " 'Powerful new technologies are becoming available that justify a major technology development effort to provide future technical options to implement a technical strategy' " (8). Though these technological capabilities are "not here now" and are only "on the horizon," they "are coming and are foreseeable" (8). Chief of these capabilities is that offered by computers and the software operating them:

It has been the incredible leaps in data processing, as much as any single area, which has fueled this explosion [of defense options]. And it is not just that we no longer need mammoth warehouses to contain the radars and computers necessary to the ABM of the 1960s. The very existence of today's and tomorrow's ability to solve complex problems on incredibly small machines, and fast, has opened up the development of our entire national technical base.

It was data processing which overcame John von Neuman's [one of the greatest mathematicians of our century] skepticism of ever making the ICBM work in the first place. It was data processing at the heart of the move to MIRVing [multiple, independently targeted warheads on a single missile]. It was data processing which tied ICBM fleets together for coordinated execution. It was data processing which has provided the ICBM accuracy necessary for preemptive strikes. And it is data processing which will be at the heart of any defense against ballistic missiles.

The plausibility of such technologies makes it imperative that we investigate the development of these defensive capabilities (8–9).

This statement addresses computer power and size and presents statements that are almost all clearly true because they deal with past and present realities. The difficulty, though, lies in whether the hopes for the future are realistic. These hopes hinge not on computing power per se, however, as we will see, but on the correspondence between software programs and the complex realities of a determined enemy trying to defeat the system at some unknown time in the future.

Next to testify was Lieutenant General James A. Abrahamson, director of the Strategic Defense Initiative Organization. Lt. Gen. Abrahamson's statement has a level, matter-of-fact tone that presents goals, aims, and intentions as practical realities. It is as though the desirability of these goals, coupled with American technical ingenuity, cannot help but succeed. Thus hopes and realities become confused in the language:

> The Battle Management/Command, Control, and Communications technology project will develop the technologies necessary to allow eventual implementation of a highly responsive, ultra reliable, survivable, endurable, and cost effective BM/C3 system for a low-leakage defense system. The BM/C3 system is expected to be quite complex and must operate reliably even in the presence of disturbances caused by nuclear effects or direct enemy attacks. This program seeks to (1) develop the tools, methods, and components necessary for development of the BM/C3 system, and (2) quantify the risk and cost of achieving such a BM/C3 system to control the complex, multi-tiered SDI system. The systems analysis project will provide overall SDI systems guidance to weapons, sensors, C3, and supporting technologies (21). . . . And I guess my experience as a technologist and as a manager with a long career in this effort is that *we indeed can produce miracles* [emphasis added]. I think that is what the shuttle program has recently shown me, and I believe that American technical muscle can, over a long period of time, do precisely what has been laid out (30).

The response of Senator John Glenn to the technological optimism of Abrahamson, Keyworth, and other SDI proponents was incisive:

> A very basic point that Dr. Keyworth made starting out was that we were not going to discuss specific technologies; not now. And yet, the whole thing rests on that.
>
> I have followed all the different types of lasers, high velocity kinetic energy weapons, particle beams, and such for I guess about 15 or 18 years now, and my big objection to this whole thing is it has not yet been invented.
>
> I have supported laser research and particle beam research for ground use and for battlefield use and so on in the thought that perhaps it might have some application in the future, perhaps in space, too. But it seemed to me, and it seemed to me ever since the President has talked about this that what we are talking about is something that has not yet been invented. While I support research, all this business of going on with proposals for deployment and such are so premature at this point that I cannot believe we are discussing layers 1 through 5 and are just assuming that the basic physics of this thing works. That is a long way from being proven (74).

Dr. Keyworth moments later responded: "Senator Glenn, I agree with virtually every word you have said" (76).

Also voicing grave concerns about technical feasibility was Dr. Sidney D. Drell, deputy director of the Stanford Linear Accelerator. Drell pointed out a fundamental problem having to do with the computers and software running the system: "The entire system, with its many hundreds of advanced sensors and interceptors and its several battle management requirements, would have to work to almost 100 percent perfection the first time it was used, although never fully tested under realistic conditions" (199).

Another incident during the hearing shows us that seemingly hard numbers can have only weak foundations reflecting more hopes than realities. Earlier in the hearing Lt. Gen. Abrahamson and Dr. Keyworth had stated that SDI would involve a multitiered system of defenses ranging from the boost phase of a missile's flight all the way to the terminal phase near its target. By having a system of five layers with 85 percent effectiveness for each layer, they claimed that the entire system would have a "leakage" rate of only .01 percent, near perfect. Though the mathematical calculation might be correct, the reality behind the numbers is flawed.

> **Senator Tsongas** [asking Dr. Keyworth]:   How did you happen to pick, in open public testimony, 85 percent as an illustration? We talked about expectations. The only figure that has been referred to was one that was 99.99 percent effective. If it [the effectiveness of each layer] were 50 percent effective, my calculation was 312 warheads would survive. How did you pick 85 as opposed to 50, for example?
>
> **Dr. Keyworth:**   Because it was a precalculated number; that, in fact, I was aware of the total kinds of numbers, and it is in some of the reporting as something that is attainable.
>
> **Sen. Tsongas:**   But we have not even discussed the physics of it. How do we arrive at 85 percent? That is a very specific number.
>
> **Mr. John Gardner** [called upon by Lt. Gen. Abrahamson to assist]:   Last summer, of course, Dr. Fletcher and the group of people, of which I was a member of the group, spent a good amount of time trying to attempt to understand what the technical potential was. In the process of doing that, a number of different concepts were identified. As those concepts were identified, some of the technical expectations of what they might be able to do were also identified.
>
> The notion, I think, Gen. Abrahamson is referring to is the notion that the technologies that we hope to be able to demonstrate hold the promise for being able to operate in this region of performance levels.

Later the discussion returned to the 85 percent figure in the testimony of Mr. Albert Carnesale, president of Harvard University and former Chief, Defensive Weapons Systems Division of the U.S. Arms Control and Disarmament Agency.

Sen. Tsongas: Where did the 85 percent figure come from?

Mr. Carnesale: I think the more interesting question is where did the five layers come from? It used to be three layers. I can imagine where both numbers came from, though I was not there at the time. What effectiveness do you need to get down to one warhead [leaking through out of a total 10,000 launched in a nuclear attack] with three layers? The answer is about 98 percent. That is too high; they would laugh us out of the room. Try 85 percent. But 85 percent will not do it with three layers. How many does it take? Five layers. So the answer is five.

Sen. Tsongas: But let us put ourselves in his shoes. Everyone is saying this thing has been oversold. Now, the only figure that was referred to at all in the entire morning [of the hearing] was the 85 percent through five layers, and that is a 99.99 percent effective system.

Is that, from a science or engineering perspective, a—not credible—but justified? Can someone with a serious technical background use a figure like that in this context legitimately?

Mr. Carnesale: Not that I know of. But more importantly . . . these are not five completely independent systems. It is wrong, mathematically, to say that the probability of leakage is the probability of penetrating through five independent systems each of whose likelihood of intercept is .85. Do they not share, for example, sensors in common?

Suppose your satellites get shot down—your sensor satellites. Let me tell you, that boost-phase system is not going to do anything, and neither is either leg of mid-course intercept. So three systems just went to zero from .85. The layers are not independent. The real defense would be far less effective.

These examples show that seemingly definite technical information can be derived from speculation and wishes and from backward reasoning that might not hold up under scrutiny. The "leakage" of .01 percent is valid only if the wished-for technologies were actually discovered or invented for each tier. It also assumes that the entire incredibly complex system worked exactly right through all its tiers the very first time it was actually used and then while under full-scale nuclear attack by a determined and technically sophisticated enemy. Otherwise, the number is practically meaningless.

In the next section we will see how additional voices arose both within and outside of government to echo these official concerns.

SDI DOCUMENTS, PRO AND CON

Still the basic question remained about the president's plan, however: Could it be done? Two opposing sides emerged to answer to this question. In the following two sections, statements in support of the potential feasibility of SDI software will be presented (Pro), followed by statements arguing that trustworthy, realistic software is not feasible (Con). Only a few representative statements can be offered here. Keep in mind that the software system to operate devices had not even been invented yet. Nevertheless, the fundamental question of software realism and trustworthiness is crucial because it is the "glue" that holds all the parts together and makes them work as a system.

Pro

Many observers have commented on the peculiarly disparate nature of the debate about SDI. Opponents of SDI nearly always cited the technological impediments to any realistic expectation that the system would do anything remotely like what was claimed for it. These included both scientists and political and defense experts.

Proponents of SDI spoke a different kind of message. "In contrast, proponents began their arguments with statements concerning moral, political, or ideological statements in order to justify the need for the radical change embodied in SDI," says Gerald Steinberg, a political scientist specializing in peace studies and conflict policy (100). They would commonly begin by describing how terrible nuclear war would be and how desirable a strong defense against such a war would be.

Proponents would not, however, directly emphasize technological feasibility. When technological matters were addressed, it was in vague or general terms celebrating past technological achievements, celebrating U.S. ingenuity and resourcefulness and celebrating a great hope for future technological advances. Steinberg explains that the celebration of past technological achievements generates a sort of patriotic fervor that distracts one from considering the realities of SDI.

More important, it also seems to presuppose the desirability of a technological fix to any and all problems. This, in turn, distracts us from the realistic understanding that technology has given us the very nuclear weapons that we are so desperate to defend ourselves against. It also deludes us into thinking that we can fully control our own destiny—all we need do is put more of our minds, energy, and money to the task. We succeeded when we put our minds to making the atomic bomb in the Manhattan Project and to landing on the moon in the Apollo program, it is typically argued. A statement by SDIO head Lt. Gen. Abrahamson is typical of this attitude: "I don't think anything in this country is technically impossible. We have a nation which indeed can produce miracles" (Congr. Hrg. 30).

The truer reality is that there are many sides to the problem of preventing nuclear war, and many parties and countries involved. A "technological fix" naturally seems attractive because it seems quick, neat, and sure. Political negotiations, though, are messy, protracted, and indeterminate. Many critics point out, however, that the appeal of SDI over foreign policy negotiations is only a superficial one, seductive though it may be. The honest reality is that people have to meet with, consider, and communicate with other people in order to work out the problems they mutually face.

Almost two years after SDI was first proposed by the president, the Office of the President issued a lengthy official statement articulating the plan only sketched two years earlier and addressing concerns raised by critics. The theme of technological optimism was reiterated but now rather tentatively:

> The President's goal, and his challenge to our scientists and engineers, is to iden-
> tify the technological problems and to find the technical solutions so that we have
> the option of using the potential of strategic defenses to provide a more effective,
> more stable means of keeping the United States and its allies secure from aggres-
> sion and coercion. . . . Recent advances in ballistic missile defense technologies . . .
> provide more than sufficient reason to believe that defensive systems could even-
> tually provide a better and more stable basis for deterrence (3).

George A. Keyworth, science advisor to the president, in April 1984 wrote a supporting article with clear ethical resonances in its title: "A Sense of Obligation—The Strategic Defense Initiative." He explained why the president proposed SDI. "He did this because he knows the technology is available, or becoming available, with which effective defense can be developed" (56). He also explains that he is not suffering under "the delusion that we can pull some incredible technological rabbit out of the hat" (62). In starting the SDI pro-gram, the president "offered our national technical community both a moral and a scientific challenge" (62).

In the fall of 1984, only a year after SDIO has been established, Keyworth wrote a 14-page technical article, "The Case for Strategic Defense." It shows the pattern of ethical and rhetorical appeals that Gerald Steinberg discussed ear-lier. Almost the entire article focuses on ethics in the sense of values and moral appeals. Among these are the desirability of a real defense against a terrible threat; the inherent rightness of being about to protect ourselves; the goodness of our resisting an evil, brutal enemy; and the good sense of learning lessons from history. Wouldn't we be morally right to do this, it argues, rather than arguing, Yes, we can do this and do it well.

Only a few pages discuss technical feasibility and then only with claims rather than concrete supporting evidence. Just as important, even the top techni-cal advisor to the president had reconceived the aim away from rendering missiles "impotent and obsolete" by building a protective shield across the United States. The aim of SDI was changed in two ways, according to Keyworth. One, it would now provide an additional sort of deterrence against an enemy's use of nuclear missiles. Recall that the president had said a primary motivation to pursue SDI was to move away from deterrence. Two, it was to provide additional leverage in arms control negotiations, which it could do even if it worked imperfectly or only with limited capability. The technical needs that SDI would have to fulfill were now much reduced, even though its feasibility was still to be determined.

Lt. Gen. Abrahamson, the first director of SDIO, naturally supported the program strongly. In a statement to Congress, he outlined the plan for SDI research and development only in general terms, however, which confuses goals with realities:

> The Battle Management/Command, Control, and Communications Technology Project will develop the technologies necessary to allow eventual implementation of a highly responsive, ultra reliable, survivable, endurable and cost effective BM/C3 system for a low-leakage defense system. The information processing capability, specifically the development of complex software packages . . . is expected to stress software development technology.

Elsewhere he explains how the SDI system is supposed to work. By this time, however, serious reservations about the feasibility of the software had been raised, which he tried to rebut. "While this topic [BMC3] must be the subject of substantial attention as the Strategic Defense Initiative proceeds, it is by no means clear that multi-phase [multi-tier] defensive systems cannot be structured to allow positive control" (11). It is always difficult to prove a negative assertion, but the negation of a negative assertion is even more problematic.

Secretary of Defense Casper Weinberger vigorously supported in his March 1984 Department of Defense report:

> Recent advances in technology offer us, for the first time in history, the opportunity to develop an effective defense against ballistic missiles. . . . Because of recent advances in technology, it is now possible to specify how these key functions of an effective ballistic missile defense could be met. . . . Computer hardware and software and signal processing in the 1960s was incapable of supporting such a multi-tiered defense battle management. Today, technological advances permit the development of effective command, control, and communication facilities (116–17).

In order to determine officially the technical feasibility of the goals of SDI, three panels of expert consultants were established by the White House. These are commonly known by the name of their chairs. The Miller panel studied policy implications of SDI, but its report is classified. The Hoffman panel explored policy implications of SDI rather than technical means. It found that a perfect or near-perfect SDI scheme such as the president stated would not be feasible. A very modest scheme building on current technologies garnered general but qualified support from the panel.

The most famous of the panels was the Defensive Technologies Study group chaired by James Fletcher, who also headed NASA at the time. Under-Secretary of Defense DeLauer introduced the report optimistically: "I am confident that our greatest asset, our people's ingenuity and creativity, will make the President's vision a reality" (i). The panel asks a simple, direct question. Given that twenty years earlier an effective defense was determined to be infeasible, the panel now asks,

"What has happened to justify another evaluation of ballistic missile defense as a basis for a major change in strategy?" Advances in defensive technologies warrant such a reevaluation. . . . Because of technological advances, the needed command, control, and communications facilities in all likelihood will be realized. (146–47)

The panel's official conclusion is:

The members of the Defensive Technologies Study Team finished their work with a sense of optimism. The technological challenges of a strategic defense initiative are great but not insurmountable. By pursuing the long-term, technically feasible research and development plan identified by the Study Team and presented in this report, the United States will reach that point where knowledgeable decisions concerning the engineering validation phase can be made with confidence.

Note that even this explicitly optimistic statement is couched in vague, conditional language that is typical of nearly all the technical statements made in support of SDI. It does not say that SDI is feasible or even that it is expected to be found feasible, only that after a good deal of research we will have the means to make a determination as to feasibility.

The overall optimism of the report was not shared by the entire panel, however. One of the panel's most renowned members, David L. Parnas, resigned from the panel before its report was issued. Parnas, often called the founding father of the field of software engineering, had serious reservations about the proposed SDI software system. Besides removing himself from the panel, however, he also issued a detailed explanation for his actions and a critique of the panel and SDI, which will be discussed later.

Con

Other people opposed SDI because of its infeasibility as well as for other reasons. A joint statement by McGeorge Bundy (former advisor on national security affairs), George Kennan (former ambassador to the Soviet Union), Robert McNamara (former Secretary of Defense), and Gerard Smith (former delegate to the Strategic Arms Limitation Talks), for example, stated flatly in the respected international journal *Foreign Affairs:* "What is centrally and fundamentally wrong with the President's objective is that it cannot be achieved. The overwhelming consensus of the nation's technical community is that in fact there is no prospect whatever that science and technology can, at any time in the next several decades, make nuclear weapons 'impotent and obsolete.' " (265)

Many technical experts voiced opposition to SDI immediately. Though the hardware was thought to be at least remotely feasible, the vitally important software was seen as fundamentally infeasible. Larry Smarr, then director of the National Center for Supercomputing Applications, said simply about SDI software: "It will not do what it was meant to do, and it will not anticipate everything that the enemy might throw at it" (*Science*, 367).

Herbert Lin, a computer scientist at MIT, reviewed prospects for SDI software in *Scientific American* in December 1985. Lin reiterates concerns cited in other sources such as the unprecedented complexity of the project, the inadequacy of fault detection, automatic programming, expert systems, and fault tolerance methods, the inability to anticipate "unknown unknowns," and the impossibility of realistic empirical testing.

Lin also cites several examples, including the sinking of the British destroyer *Sheffield* in the Falkland Islands war by an Argentine Exocet missile. The destroyer's computerized defenses had been programmed to disregard Exocet missiles because they were included in the British arsenal, too. The program did not correspond to reality. A second example is the first operational test of the Navy's Aegis computerized defense system on the new cruiser *Ticonderoga*. The system failed to shoot down 6 out of 16 targets due to software faults. From his analysis of the inherent challenges of the project and from many examples of software errors in complex systems, Lin concludes that the SDI software project as described in the president's speech is impossible.

Another important technical voice opposing Star Wars was the Union of Concerned Scientists, a professional organization numbering in the thousands. In 1984 the union published a book-length critique of SDI. Among its concerns was that the fundamental validity of the algorithms running the computer could never be proven to encompass all conceivable threats. In 1986 the union published another book critical of Star Wars. It identified three "weak links" in the SDI program, which by then had evolved from a program of complete protection for the entire nation to a limited defense for selected potential targets meant only to deter an attack rather than to nullify one. The book succinctly describes how "technological optimism" underlay practically all the stellar yet unrealistic projections for how effectively Star Wars would perform. This technological optimism was, they said, unwarranted due to the failure of the analogy between historical technological accomplishments and the SDI proposal.

An entire chapter was devoted to software. It concludes:

> The SDI is inherently dependent on its computer system, without which any collection of exotic weapons, sensors, and strategies would be useless. The required computer system would be far more complex than any previous computerized weapon system and would require software well beyond the current state of the art. Given the magnitude of this software engineering challenge, there is a significant chance that the development effort would simply fail to produce a deployable system at all. . . . Optimists can believe that a system could be deployed that would work if it were needed. Because operational testing is impossible, however, the issue could never be settled with certainty. The reliability of the system would always be in doubt (106).

Parnas

David L. Parnas is often called the father of software engineering and has extensive experience in designing military software systems. He resigned in protest from the Fletcher panel and in a subsequent statement explained his reasons for resigning. This statement (Figure 7.2) is a remarkable piece of technical communication. Not only does it deal with a very sophisticated technical topic in easy-to-understand language, but it also takes a clear, explicit ethical stand in support of a technical position. It melds technical information with ethical concerns in a way that each supports the other. For Parnas, the technical is ethical and the ethical is technical. It is the most important statement ever made critical of SDI software, stating objections that remain valid to this day.

Parnas's statement on SDI is an exemplary model of clear, effective, and ethical technical communication in many ways, discussed next. The complete statement spans about ten pages, so only a few excerpts have been presented here.

Parnas's statement is highly technical yet clear. It deals with fundamental issues about the theory and practicability of complex software systems, but operates at the level of first principles, basic concepts, and fundamental knowledge. It is therefore general but not vague; instead, it is direct, incisive, and entirely clear. It has consistent focus and a clear, coherent development of its arguments, all of which deal with crucial points.

Many of the statements of SDI supporters, on the other hand, are very circuitous. They present a great deal of background information and lengthy explanation of detailed processes or devices but devote little space to clearly stating the significance or realistic feasibility of what is being discussed. This leaves the reader feeling swamped by technical details and feeling expected to acquiesce to a summary conclusion out of exhaustion, confusion, or intimidation. Parnas does the opposite.

His statement accommodates its audience very effectively. It communicates highly technical concepts in language that it accessible to any educated, nontechnical reader. A good deal of everyday technical communication in many other fields has to accomplish this same task in order to be effective and usable. Its criticisms are clearly explained and motivated by values that are meaningful to the audience. It thus focuses on the task of clear communication and effective persuasion of its audience, rather than on political posturing or esoteric objections that are understandable by only a few.

Statements of SDI supporters, on the other hand, often inundate the reader with a flood of technical details and enticing projections, presenting an image of the writer's technical expertise that intimidates rather than accommodates.

Parnas's statement takes a definite ethical stance that is clearly articulated, firmly defined, and appropriate for the context. Parnas repeatedly insists that the public should not be misled about security from SDI and that they not be misled about the likely effectiveness of the vast sums spent on SDI. He also takes care to explain that his conclusions are expert professional judgments from someone with military software experience. (At the time, many other scientists avoided defense contracts because of their antimilitarist politics.) His statements, he points out, have to do with the problems inherent in SDI itself, not with any personal objection to military research that he might harbor (which in fact he does not).

Software Aspects of Strategic Defense Systems

The essays that constitute this report were written to organize my thoughts on these topics and were submitted to SDIO with my resignation. . . .

My conclusions are not based on political or policy judgments. Unlike many other academic critics of the SDI effort, I have not, in the past, objected to defense efforts or defense-sponsored research. I have been deeply involved in such research and have consulted extensively on defense projects. My conclusions are based on more than 20 years of research on software engineering, including more than 8 years of work on real-time software used in military aircraft. They are based on familiarity with both operational military software and computer science research. My conclusions are based on characteristics peculiar to this effort, not objections to weapons development in general. . . .

I am publishing the papers that accompanied my letter of resignation so that interested people can understand why many computer scientists believe that systems of the sort being considered for SDIO cannot be built. . . .

The lay public, familiar with only a few incidents of software failure, may regard them as exceptions caused by inept programmers. Those of us who are software professionals know better; the most competent programmers in the world cannot avoid such problems. . . .

All of the cost estimates indicate that this will be the most massive software project ever attempted. The system has numerous technical characteristics that will make it more difficult than previous systems, independent of size. Because of the extreme demands on the system and our inability to test it, we will never be able to believe, with any confidence, that we have succeeded. Nuclear war will remain a potent threat. . . .

Programming is a trial and error craft. People write programs without any expectation that they will be right the first time. They spend at least as much time testing and correcting errors as they spent writing the initial program. Large concerns have separate groups of testers to do quality assurance. Programmers cannot be trusted to test their own programs adequately. Software is released for us, not when it is known to be correct, but when the rate of discovering new errors slows down to one that management considers acceptable. . . .

FIGURE 7.2   Parnas's Critique of SDI Software These are excerpts from David L. Parnas's famous published critique of claims made about proposed SDI software. It is highly technical yet also employs rhetorical appeals dealing with ethical and value concerns. Basically it explains that the software could not be absolutely trusted to do what would be needed of it due to issues of a fundamental and irresolvable nature.

Originally published in *American Scientist* and then reprinted in *Communications of the ACM* (Association for Computing Machinery). By permission of *American Scientist*.

Although I believe that further research on software engineering methods can lead to substantial improvements in our ability to build large real-time software systems, this work will not overcome the difficulties inherent in the plans for battle-management computing for SDI. Software engineering methods do not eliminate errors. They do not eliminate the basic differences between software technology and other areas of engineering. They do not eliminate the need for extensive testing under field conditions or the need for opportunities to revise the system while it is in use. Most important, we have learned that the successful application of these methods depends on experience accumulated while building and maintaining similar systems. There is no body of experience for SDI battle management. . . .

I believe that the claims that have been made for automatic programming systems are greatly exaggerated. Automatic programming in a way that is substantially different from what we do today is not likely to become a practical tool for real-time systems like the SDI battle management system. Moreover, one of the basic problems with SDI is that we do not have the information to write specifications that we can trust. . . .

Even if size were not a problem, the lack of specifications would make the notion of a formal proof meaningless. If we wrote a formal specification for the software, we would have no way of proving that a program that satisfied the specification would actually do what we expected it to do. The specification itself might be wrong or incomplete.

The SDIO panel on battle management computing contains not one person who has built actual battle management software. It contains no expert on trajectory computations, pattern recognition, or other areas critical to this problem. All of its members stand to profit from continuation of the program.

In several discussions of this problem, I have found people telling me that they knew the SDIO software could not be built but felt the project should continue because it might fund some good research. . . . There is an obvious moral problem raised by this position. . . . There is no justification for continuing with the pretense that the SDI battle management software can be built just to obtain funding for otherwise worthwhile programs.

FIGURE 7.2   Continued

Parnas's chief point is that no software system could possibly be developed along the lines required by SDI that would be "trustworthy." Trustworthy in this context means that we can have confidence that it will do what it is supposed to do. The software system might be hampered by faults or errors within itself. It could also break down due to reliability problems such as communications breaks, power outages, or synchronization errors (such as delayed the launch of the very first space shuttle). It might also be unable to grapple with the many counteracting strategies and tactics that an enemy might present, many of which could not have been foreseen.

More fundamentally, many of these potential difficulties could not be completely anticipated and debugged beforehand simply because we can never be sure of what we do not know or understand. Still more importantly, we can never be sure that all possible difficulties have been revealed through our prior simulations because no fully legitimate simulation regimen can replicate any and all realistic battle circumstances. The only thing that can truly simulate real battle is real battle itself, but real nuclear battle would be fought only once, in a matter of hours, leaving no room for debugging or real-time error correction. Thus, Parnas concludes, it is impossible that we could ever develop a completely trustworthy software system for SDI. Thus the task before the panel is fundamentally impossible.

An additional important ethical component of Parnas's explanation for his leaving the panel is that he would not collude with the panel in the pretense of pursuing the goal of SDI. The panel members could, for example, continue to receive the defense funds just in order to pursue their own interests even though they might really believe SDI to be unworkable. Parnas is clear about the need for truthfulness and for frank, open discussion in public about technical and strategic realities.

In his final point, that SDIO is not an efficient way to fund research on subsystems even if the total system is unworkable, he explicitly identifies this as a "moral issue" under which the public should be told what they can reasonably expect for the vast amount of money they would be spending on SDI.

Parnas's statement articulates its ethos. "Ethos" is a rhetorical term referring to believability on the basis of perceived character. Aristotle says that at times ethos is even more important than any evidence or reasoned argument one might offer. Parnas's ethos is crucial to the persuasiveness of his points. As for himself, he says that at that time he had more than 20 years of experience in software engineering and over 8 years of work on real-time software in military aircraft. Indeed, he then was still serving the U.S. Navy Research Laboratory as head of its Software Cost Reduction project as the country's foremost expert on software engineering.

Parnas had something to lose in resigning, too. Though he was being paid about $1,000 per day (very big money in 1984) to serve on the panel, he resigned after only a short time instead of postponing this action for greater profit. He might also have lost later funding prospects had he earned a reputation as a troublemaker. His selfless gesture of resigning reinforces his credibility, for he had nothing to gain by it and a good deal to lose.

His most incisive statement about credibility based on personal character, however, has not to do with himself but with the rest of the panel. "The SDIO panel on battle-management computing contains not one person who has built actual battle-management software [excluding himself]. It contains no experts on trajectory computations, pattern recognition, or other areas critical to this problem. All of its members stand to profit from continuation of the program." He, on the other hand, has all these qualifications.

Parnas's statement shook the national consciousness. Though other software engineers and computer experts have echoed Parnas and generally affirmed his arguments, he himself has served as the rallying point for the critical opposition. His clear, calm, cogent arguments brought an air of common sense to the SDI debate that cut through a cloud of wishful thinking and political brow beating that had led to the funding of research without clear goals or realistic expectations. His statement was popular also because it affirms the layperson's intuitive sense that a computer system deals only mechanically with the information with which we supply it, which might have little correspondence to reality.

A few years after Parnas resigned from the panel, in a separate statement he further explained the sense of professional ethical responsibility that motivated his action. What he says regarding himself as a software engineering professional can apply as well to all technical communication professionals:

> My decision to resign from the panel was consistent with long-held views about the individual responsibility of a professional. I believe this responsibility goes beyond the obligation to satisfy the short-term demands of the immediate employer. As a professional:
>
> - I am responsible for my own actions and cannot rely on any external authority to make my decisions for me.
> - I cannot ignore ethical and moral issues. I must devote some of my energy to deciding whether the task that I have been given is of benefit to society.
> - I must make sure that I am solving the real problem, not simply providing short-term satisfaction to my supervisor.
>
> It is not necessary for computer scientists to take a political position; they need only be true to their professional responsibilities. If the public were aware of the technical facts, if they knew how unlikely it is that such a shield would be effective, public support would evaporate. We do not need to tell the public not to build SDI. We only need to help them understand why it will never be an effective and trustworthy shield (in Johnson, 15–25).

Though Parnas is talking about his values, he does not call this "politics." Rather than voicing his politics, he is only insisting on the truth and on making the truth as clear and open to the public as possible. This is "politics" perhaps, but only in the best sense, by which the overall public good is pursued over narrow self-interest.

Note that Parnas sees no inconsistency between his professional responsibilities and his public, civic responsibilities. Neither does he see any inconsistency between his professional responsibilities and his personal ones. To him they are all the same, as though to be ethical means all these things at once. This view is actually the same as that of all but one of the ethical perspectives we have examined in this book, though each had a somewhat different way of articulating the point.

For Plato, the personal good is of prime importance; when this is attended to, it naturally follows that the public good will be pursued as well. Similarly for Aristotle; if the personal good in the form of virtue is pursued, then civic good will follow as well. Though, to be sure, Aristotle does at times posit a conflict between personal and civic "goods" and between personal and professional obligations, these are often fine distinctions that in practice would arise only very rarely. For Kant, the personal good necessarily has to be equivalent to the general civic good in the sense that we would all wish to treat and be treated the same. Our innate sense of rightness as arrived at rationally would guide our professional activities in relation to others in ways such as we would wish to be treated ourselves. For most feminist ethicists, the identity between the personal, the professional, and the political is a basic principle. They would insist on authenticity, sincerity, and honesty. In addition, an ethic of care perspective would of course insist as well on a caring attitude toward all and a concern for the give-and-take of ongoing communication. A similar caring concern for the public good through honest, open discussion is found in Parnas's ethics, too.

Only in the case of utilitarianism does the consonance among our perspectives break down. Though with utilitarianism the personal good need not always differ from the public good, it is assumed that it oftentimes will; this in fact is one of the distinguishing features of utilitarianism. Likewise, the personal good might differ from the professional good or the public good.

In the case of SDI and Parnas, in fact, some defense researchers and software engineering professionals criticized Parnas specifically for not trying to protect the image of his profession. Not only was he making the panel itself appear inadequate and perhaps unethical, but he was making the entire defense research establishment appear selfish and ineffective. At the same time, his statements might result in reduced funding for lucrative research projects for some of the researchers. Parnas himself cites several instances in which colleagues wished his silence so that their funding sources would not dry up. "This may be a businesslike attitude," he says, "but it is not a professional one. It misleads the government into wasting taxpayer's money" (in Johnson, 24). It is no coincidence, of course, that utilitarianism, of all the major historical ethical perspectives, is the one most often criticized as the least genuinely ethical one.

For his groundbreaking ethical and technical critiques of SDI, Parnas was awarded the first Norbert Weiner Award for Social and Professional Responsibility by the organization Computer Professionals for Social Responsibility in 1987.

STAR WARS BOYCOTT PLEDGE

The Star Wars debate was extraordinary in the annals of technical and scientific history in many ways but chiefly in stirring among technicians and scientists their sense of ethical responsibility. No other technical and scientific issue in modern times—except for the development and proliferation of nuclear weapons themselves—has so stirred this community. This sense of ethical responsibility was coupled with a civic responsibility and burst forth dramatically in the form of the public Star Wars Boycott Pledge.

In the early days of the Strategic Defense Initiative, the Department of Defense had identified academic scientists and technical experts for special efforts to elicit their support of the program. Nonacademic scientists, on the other hand, had an obvious self-interest in procuring defense contracts and so would not be considered a disinterested voice in the debate. Despite the Department of Defense's efforts, many scientists and technical experts opposed SDI and expressed their opposition clearly. At AT&T's Bell Laboratories, for instance, 1,600 technical experts signed a petition opposing SDI. Extensive surveys by the National Academy of Science and the Union of Concerned Scientists found broad opposition to SDI among most scientists and technical experts.

The most significant expression of opposition came in the form of a petition originating in 1985 at the University of Illinois and Cornell University, the Star Wars Boycott Pledge (Figure 7.3). It was a short, direct statement in which signers

---

Star Wars Boycott Pledge

We, the undersigned scientists and engineers, believe that the Strategic Defense Initiative (SDI) program (commonly known as Star Wars) is ill-conceived and dangerous. Anti-ballistic-missile defense of sufficient reliability to defend the population of the United States against a Soviet attack is not technically feasible. A system of more limited capability will only serve to escalate the nuclear arms race by encouraging the development of both additional offensive overkill and an all-out competition in anti-ballistic-missile weapons. The program will jeopardize existing arms control agreements and make arms control negotiation even more difficult than it is at present. The program is a step toward the type of weapons and strategy likely to trigger a nuclear holocaust. For these reasons, we believe that the SDI program represents, not an advance toward genuine security, but rather a major step backwards.

Accordingly, as working scientists and engineers, we pledge neither to solicit nor accept SDI funds, and we encourage others to join us in this refusal. We hope together to persuade the public and Congress not to support this deeply misguided and dangerous program.

FIGURE 7.3   Star Wars Boycott Pledge This statement was circulated in 1985 through 1986 as a pledge not to accept funding related to Star Wars. It was signed by thousand of scientists, researchers, and engineers as well as professors and graduate students at almost all major research universities, including fifteen Nobel laureates. The scope of its support and depth of its concern are almost unprecedented in the technical and scientific community. It deals with highly technical subject matter yet from a clear and firm ethical basis.

pledged "neither to solicit nor accept SDI funds." The signers were of course communicating about a technical matter in this document but also about political and ethical matters. It sprang from ethical responsibility, a sense of doing the right thing.

Keep in mind that SDI was to be the most massively funded military program ever undertaken. Academic researchers, though drawing regular salaries, often depend on grants and contracts from the federal government, especially the Department of Defense. Signing the pledge was a selfless gesture that could do real damage to their reputations and livelihoods. Despite these possible negative consequences, within one year the pledge had been signed by over 3,700 scientists, engineering professors, and researchers at over 110 academic institutions. This included fifteen Nobel Prize winners and over 50 percent of the faculty of the top twenty U.S. physics departments. It was a major repudiation of the technical and scientific claims that were being made about SDI and of the policy implications of the program.

Note that the document blurs the distinction between the technical and the political, just as in our society the strong influence of science and technology brings about powerful political effects. The defense industry, for example, influences the government to support its scientific and technical activities. At the same time, the defense industry supports the government by developing the technical means to achieve political ends. The "smart" weapons of the Persian Gulf war, for example, allowed our forces to attack without risking face-to-face combat.

Michael Nusbaumer, a sociologist, and his colleagues later researched the motives of the signers of the pledge. His research shows, among other things, that those with the greater technical and scientific knowledge among pledgers were more likely to object to SDI on ethical grounds. In general the pledgers felt that ethical concerns were of great importance in claims made about the technical feasibility and desirability of SDI. Their conclusion applies to both science and technology: "Scientists of all types are increasingly aware of and are being held responsible for their science and its relationship to larger social and political issues" (385). These scientists and technical experts enacted their ethics on this highly technical matter.

Patriot: Small-Scale SDI

Star Wars is still with us in many ways. The less exotic hardware continues to be developed, some of which will be deployed soon. The Ballistic Missile Defense Organization (BMDO) continues to develop missiles to intercept incoming warheads, for example. One part of this program involves using new variants of the Patriot missile to intercept tactical missiles (bigger, slower, nearer, and easier targets than intercontinental ballistic missiles). The Patriot system has been touted as a rudimentary form of Star Wars.

Reports about the performance of the Patriot missile system in the Persian Gulf War can provide an important lesson for us, however. They show how problematic technical communications can be even about existing concrete weapons. One would think that in communicating a description and evaluation of a limited number of concrete objects and actual events in the recent past, there would be little room for misunderstanding or for ethically questionable representations. Experience shows otherwise, however. Values, assumptions, and expectations play powerful roles in shaping even hard technical information.

Most of us can recall dramatic images from the Persian Gulf showing Patriot missiles arcing through the night sky and then exploding like fireworks against incoming Iraqi Scud missiles. Recall how proud many people felt about this marvelous technology and how grateful many felt that it was turned to the protection of innocent civilians.

Military leaders proclaimed the Patriot highly successful. Proponents of SDI especially delighted in the performance of this complex system, which seemed to demonstrate that a missile could shoot down another missile in real wartime circumstances. This technology needed only to be upscaled to allow us to intercept strategic, long-range ballistic missiles, it was said.

Only well after the war were the data and conclusions of these early reports scrutinized. The seriousness of the conflicting representations and of the significance of the Patriot system for projections about the performance of even more complex systems planned for the future led to a congressional investigation to pin down the reality behind the representations. The congressional report says that the official estimates by the U.S. Army of the successful destruction of Scuds by Patriots began at 100 percent during the war, then fell to 96 percent, then 80 percent, then 70 percent, then 52 percent. Independent estimates and the congressional committee's conclusions are worse, in the neighborhood of 9 percent or even less. Most of the time, that is, the Patriot did not destroy its target and frequently did not even get close.

Patriot thus produced a poor first showing with a high leakage rate, hardly encouraging for those trying to draw an analogy to SDI. The Patriot batteries also were not under concentrated attack. The congressional investigation did not find any unethical activities, we should note: "It is probable that many of the individuals . . . including the President of the United States and General H. Norman Schwarzkopf, were not aware at the time that the claims of success were false" (1).

The congressional report emphasizes the ethical responsibility at stake regarding the Patriot information:

> American soldiers' lives could be unnecessarily endangered if they are deployed in future conflicts based on inaccurate assessments of the Patriot's capabilities. They may depend on Patriot battalions destroying almost all of the enemy missiles, as the Army now claims, when the actual defensive capabilities may mean that it could actually miss almost all of the threatening missiles. (1)

The purpose of citing this congressional report is only to show that seemingly incontrovertible factual information should not be accepted uncritically at face value. What the data precisely are, how they are arrived at, how they might be verified, and what they mean are all subject to the influence of various factors that shape the final representation in important ways. Thus apparently definite technical information can actually be much less definite than it seems.

Our responsibility as technical communicators is to try to ensure as best we can that our representation of information corresponds to the reality. Though we usually do not originate the information, a prudent skepticism and caution in representing technical information would be appropriate.

Technical Claims about Air Operations

Another important illustration about the validity of claims made in technical documents about the capabilities of technical devices can be found in the GAO report *Operation Desert Storm: Evaluation of the Air Campaign* in June 1997. GAO examined in detail the validity of claims made by various sources about the air operations during Desert Storm, particularly about the effectiveness and suitability of new weapons systems such as precision guided munitions ("smart bombs") and the F-117 stealth attack aircraft. Figure 7.4 shows one of the tables in the report and dramatically illustrates the difference between claims made about the effectiveness and suitability of various weapons systems during the war versus the findings of the GAO about their actual performance during the war. Some of these claims were made by the Department of Defense and some by the manufacturers, but in general the two entities were making the same substantive claims. For example, the claim that the LANTIRN infrared imaging system was able to "locate and attack targets at night and under other conditions of poor visibility" was shown not to be true unless the system was employed *"below* clouds and weather." Of course, clouds would seem to be a rather common condition of "poor visibility" that readers would assume to be encompassed in the manufacturer's statement. Most readers would also assume that the wording and tone of the claim refers to *any and all* conditions of poor visibility, though the claim holds true only for certain selected conditions of poor visibility. As in the case of the Patriot missile system, we need to keep in mind that these claims and technical documents are not dealing with something as innocuous as a home appliance. Instead they are loaded with real, practical significance because the success of military operations and ultimately lives are at stake.

ETHICAL APPRAISAL

We have reviewed the great debate over the Strategic Defense Initiative, potentially one of the most important and most expensive military systems ever devised. We began with President Reagan's dramatic speech sketching the goals of a plan that was supposed to provide security from the threat of nuclear ballistic missile attack.

TABLE III.16: Manufacturers' Statements About Product Performance Compared to GAO Findings

MANUFACTURER	PRODUCT	STATEMENT	FINDING
General Dynamics	F-16	No matter what the mission, air-to-air, air-to-ground. No matter what the weather, day or night. The F-16 is the premier dogfighter."[a]	The F-16's delivery of precision air-to-ground munitions, such as Maverick, was impaired, and sometimes made impossible, by clouds, haze, humidity, smoke, and dust. Only less accurate unguided munitions could be employed in adverse weather using radar.
Grumman	A-6E F-11	"A-6s . . . [were] detecting, identifying, tracking, and destroying targets in any weather, day or night."[b]	The A-6E FLIR's ability to detect and identify targets was limited by clouds, haze, humidity, smoke, and dust; the laser designator's ability to track targets was similarly limited. Only less accurate unguided munitions could be employed in adverse weather using radar.
Lockheed		Achieved "80 percent direct hits."[c]	The hit rate was between 55 and 80 percent; the probability of bomb release was only 75 percent; thus, the probability of a hit during a scheduled F-117 mission was between 41 and 60 percent.

(continued)

FIGURE 7.4   GAO Table Comparing Claims vs. Performance This table compares technical claims made by the manufacturer of aircraft, weapons, and other technical devices used in Desert Storm in the conflict with Iraq, against the GAO determination of the actual performance demonstrated in the conflict. According to the GAO, "the manufacturers made public statements about the performance of their products in Desert Storm that are not fully supported" (145). The impression communicated to the public and to various governmental agencies and people was basically incorrect or misleading.

From GAO/NSIAD-97-134 *Operation Desert Storm: Evaluation of the Air Campaign*, June 1997, pp. 144–145.

MANUFACTURER	PRODUCT	STATEMENT	FINDING
		The "only aircraft to attack heavily defended downtown Baghdad."[c]	Other types of aircraft frequently attacked targets in the equally heavily defended metropolitan area; the Baghdad region was as heavily defended as downtown.
		"During the first night, 30 F-117s struck 37 high-value targets, inflicting damage that collapsed Saddam Hussein's air defense system and all but eliminated Iraq's ability to wage coordinated war."[d] "On Day 1 of the war, only 36 Stealth Fighters (less than 2.5% of the coalition's tactical assets) were in the Gulf theater, yet they attacked 31% of the 17 January targets."[d] "The F-117 reinstated the element of surprise."[c]	On the first night, 21 of the 37 high-value targets to which F-117s were tasked were reported hit; of these, the F-117s missed 40 percent of their strategic air defense targets. BDA on 11 of the F-117 SAD targets confirmed only 2 complete kills. Numerous aircraft, other than the F-117, were involved in suppressing the Iraqi IADS, which did not show a marked falloff in aircraft kills until day 5.
			The 2.5-percent claim is based on a comparison of the F-117s to all deployed aircraft, including those incapable of dropping bombs. The F-117s represented 32 percent of U.S. aircraft capable of delivering LGBs with warheads designed to penetrate hardened targets. F-117s were tasked against 35 percent of the first-day strategic targets.

(continued)

FIGURE 7.4 Continued

			Other nonstealthy aircraft also achieved surprise. Stealth characteristics did not ensure surprise for all F-117 strikes; modifications in tactics in the use of support aircraft were required.
Martin Marietta	LANTIRN	Can "locate and attack targets at night and under other conditions of poor visibility using low-level, high speed tactics."[e]	LANTIRN can be employed below clouds and weather; however, its ability to find and designate targets through clouds, haze, smoke, dust, and humidity ranges from limited to no capacity at all.
McDonnell Douglas	F-15E	An "all weather" attack aircraft.[f]	The ability of the F-15E using LANTIRN to detect and identify targets through clouds, haze, humidity, smoke, and dust was very limited; the laser designator's ability to track targets was similarly limited. Only less accurate unguided munitions could be employed in adverse weather using radar.
	TLAM C/D cruise missile	"Can be launched . . . in any weather."[g]	TLAM's weather limitation occurs not so much at the launch point but in the target area where the optical [DELETED].

(continued)

FIGURE 7.4   Continued

MANUFACTURER	PRODUCT	STATEMENT	FINDING
		"Incredible accuracy"; "one of the most accurate weapons in the world today."[g]	From [DELETED] percent of the TLAMs reached their intended aimpoints, with only [DELETED] percent actually hitting the target. It is impossible to assess actual damage incurred only by TLAMs.
Northrop	ALQ-135 jammer for F-15E	"Proved itself by jamming enemy threat radars"; was able "to function in virtually any hostile environment."[a]	[DELETED]
Texas Instruments	Paveway guidance for LGBs	"Employable" in "poor weather/visibility" conditions.[h]	Clouds, smoke, dust, and haze impose serious limitations on laser guidance by disrupting laser beam.
		"TI Paveway III: one target, one bomb."[a]	Our analysis of a selected sample of targets found that no single aimpoint was struck by one LGB—the average was 4, the maximum was 10.
		"LGBs accounted for only 5% of the total ordnance. But Paveway accounted for nearly 50%" of targets destroyed.[a]	Data were not compiled that would permit a determination of what percentage of targets were destroyed by any munition type.

[a]From a company advertisement in *Aviation Week and Space Technology*, (1991).
[b]Grumman Annual Report, 1991, p. 12.
[c]Lockheed briefing for GAO.
[d]From *Lockheed Horizons*, "We Own the Night," Issue 30 (1992), p. 55, 57.
[e]Martin Marietta, 10-K Report to the Securities and Exchange Commission, 1992, p. 14.
[f]McDonnell-Douglas, "Performance of MCAIR Combat Aircraft in Operation Desert Storm," brochure.
[g]McDonnell-Douglas, "Tomahawk: A Total Weapon System," brochure.
[h]Texas Instruments, "Paveway III: Laser-Guided Weapons," brochure, 1992.

FIGURE 7.4 Continued

As time went on, opposition to the plan increased on technical grounds having to do with feasibility of many features of the plan, especially its software. Some concerns were expressed even within the technical reports of the official panels. Beginning with the dramatic resignation by David L. Parnas from one of the official panels, many scientists and technical experts outside the Reagan administration began to express strong opposition on the grounds of infeasibility and lack of dependable realism and trustworthiness of the software system. In addition, technological optimism based on indefinite hopes for yet-to-be-invented new technologies-was found to be unjustified.

We can learn a great deal about ethics in technical communication from this remarkable debate and the documents we have examined. Many of the statements in support of SDI show how communications about technical matters can be expressed in needlessly vague language or with few specifics presented. The effect of this is, on the one hand, to present the audience with enough information to stimulate a strong interest in the topic without, on the other hand, allowing potential opponents any firm statements to challenge.

The following appraisals refer only to the statements of SDI proponents.

Aristotle. From the Aristotelian perspective, it is unclear whether the statements of supporters could be characterized as representing a virtuous persona. At the surface level, they adopt an ethical stance toward strategic security and favor ethical rightness (avoiding violence) over sheer practical effectiveness (the nuclear stalemate through the threat of mutual violence). On the other hand, they mask or deny any suggestion that there is little realistic feasibility to claims they are making for the overall system they are proposing.

Kant. From a Kantian perspective, these statements are problematic, also. If the statements tried to reflect as realistic what really was known not to be realistic, and the communicators did not make reasonable efforts to make an honest appraisal of feasibility, most people would agree that they would not wish to be treated this way. Therefore, the statements are questionable ethically. If, on the other hand, the statements were deliberately intended to leave a false impression on the American public, most people would agree that they would not wish to be treated this way too. To be sure, national security interests of which we might be unaware could have led these people to assert national interest over strict candor and might have been right to do so. We just do not know.

Utilitarianism. From a utilitarian perspective, the calculated weighing of national interests against individual interests such as just mentioned would have been entirely acceptable ethically. If, on the other hand, the statements were not legitimately aimed at intended effect on the Soviet Union, then they are just urging the American people and government to fund a massively expensive defense project that was doomed to be ineffective. In that case, the calculation of cost versus benefit, adjusted for the numbers of people involved, would show these statements to be unethical.

Feminist Perspective and Ethic of Care.   From a feminist ethical perspective, an appraisal is somewhat complicated. Most feminist ethicists are firmly opposed to violence and militarism in general. Thus statements calling for the means to nullify the potential violent aggression of the Soviet Union might well be approved of. For the United States itself, furthermore, avoiding a violent counterattack on our part would be ethically desirable, too. An ethic of care that of course would insist on a caring concern for the American people, and implicitly for all the world's people, would seem to be supported by these statements, too, taken at face value.

On the other hand, seen in its larger social–political context, these statements seem to be aimed at cutting off dialog with knowledgeable technical experts, which would not be ethical. At the same time, they appear to be making a needlessly authoritarian gesture in cutting off communication and, in effect, disempowering dissenting voices, which would not be ethical from a feminist perspective. From the perspective of an ethic of care, one would wonder too about the enormous amount of federal money expended on what is still a military system, albeit a defensive one. In the social context of continuing poverty, weak education, poor health care, and racial and gender inequities for so many people, one could readily suggest more socially constructive uses for that same vast amount of money.

The resignation statement of David L. Parnas is also a technical communication though of a somewhat unusual nature and so deserves its own separate treatment. It deals with technical matters at a sophisticated conceptual level but is addressed to both a specialist, technical audience and to a public, nontechnical audience. Though it is highly technical, it is also highly ethical. It reflects a person who takes his ethical responsibilities seriously at all levels—public, professional, and personal. It does this by addressing the audience with utmost honesty, with an even-handed and well-informed consideration of everything then known about the subject, and with a caring concern for the civic good. It is also clear about stating Parnas's own ethical perspective, which is a very civic-minded one. Furthermore, it offers an ethical critique beyond what most people would expect. Besides critiquing the software engineering issue alone, Parnas also goes on to educate the public by explaining why other researchers and technical experts might not have spoken out frankly on the subject before. Furthermore, he demonstrates his own ethical rectitude by denying himself the lucrative compensation he would have received had he remained on the panel, while subjecting himself to the possibility of real financial harm from being professionally blackballed.

From an Aristotelian perspective, Parnas wrote and acted in a way that sought the true, good, and right in this matter. Even most of his colleagues, to whom he directed a rather embarrassing public reproof, have admitted the basic technical soundness and ethical rightness of his statements.

From a Kantian perspective, the appraisal is equally clear. Parnas is treating his audience as he would wish to be treated himself, the bedrock of the Kantian perspective. He sees his responsibility as revealing the full truth on the matter, then trusting the public to judge the matter for themselves, again treating his audience as we imagine he would want to be treated himself. There is a sense of dignity in his relations toward us by which both he and we are dignified. In addition, he is acting out of a sense of duty rather than self-interest; indeed, he is opening himself to a great deal of potential difficulties. But his sense of duty prevails; as he explained, he felt he could not have done otherwise.

From a utilitarian perspective, his statements accomplished the benefit of avoiding fruitless expenditures (costs) on a program that could not but fail. There is no suggestion that Parnas had ulterior motives, so the only utilitarian benefits would accrue to the public and his profession.

From a feminist ethical perspective, the frank openness toward the public would be applauded, as would his rhetorical stance to educate rather than just to win approval from the audience. He affirmed the intelligence and worth of his audience and their right and power to make important decisions for themselves. From the perspective of an ethic of care, he showed a keen sense of caring concern for the public in ensuring they did not feel a false sense of security and that they get what they pay for.

Parnas's technical analysis and clear statements set a benchmark for ethical responsibility in the field of software engineering. Other observations critical of SDI by other software experts echoed the general thrust of his conclusions but usually were less strongly worded and less categorically argued. Yet nearly all his colleagues agree that the root problems that Parnas raised remain unsolved.

In assessing his ethical stance, we should recognize, too, what Parnas could have done differently. He could have said and done nothing. He could have made a more hedging and qualified statement simply calling for further research to clarify the prospects for SDI software. He could have simply gone along with the consensus of the rest of the Fletcher panel and continued to draw his earnings while allowing the unfeasibility of the SDI software project to reveal itself naturally as time went on. He did none of these and instead took the more difficult but more ethically sound route. In sum, Parnas's statement is an exemplary model of ethical technical communication.

CONCLUSION

The Strategic Defense Initiative had a laudable goal; no one can deny that. Knowledgeable critics have convincingly shown, however, that in anything like its original form, it is unworkable and could never have been trusted to do what was expected of it. There are many reasons for SDI being unworkable, the most fundamental being that the software would not work as needed. Our examination shows that serious fundamental problems about creating and using the SDI software were at first glossed over, but later criticisms became so numerous and compelling that they could not be denied.

Appealing but unrealistic claims for technical abilities of the defense establishment should show us how powerfully compelling can be our concern for basic security. This concern is so strong that it can cloud our judgment on even highly technical matters. All technical reports, we should keep in mind, involve technical information in some social or organizational context of needs and purposes. These needs and purposes not only function as the impetus behind most of our technological endeavors, but they can also shape our representations of technical information. Our ethical responsibility as communicators is to make sure that our hopes and wants do not cloud our claims about our technical abilities.

Topics for Papers and Discussion

1. This chapter reprints only excerpts from David L. Parnas's famous critique of the software proposed for SDI, "Software Aspects of Strategic Defense Systems," which also explained his ethical and technical motives for leaving one of the first panels of experts commissioned to advance the SDI program. His complete statement was originally printed in *American Scientist*, then reprinted in *Communications of the ACM (Association for Computing Machinery)* one of the most important professional organizations for computing engineers.

Read Parnas's complete statement and write a report analyzing in detail its ethical values and describing how ethical and technical expertise go hand-in-hand in this technical document. Read one or more secondary sources to bolster your comments or Parnas's later remarks in Johnson.

2. The influence of subjective values in shaping apparently objective technical information is a theme running throughout this book. One example is the credibility crisis or "flap" over claims about X-ray laser technology. X-ray lasers were, at one point in the history of SDI, meant to be the ultimate answer to the problem of how to destroy missiles in their boost phase. For many reasons, this much-heralded approach was abandoned. One of the reasons is that it would utilize a nuclear detonation as a power source in each of its devices, even though the aim of SDI was to make nuclear weapons impotent and obsolete. Another reason is sheer technical infeasibility. Different authoritative experts voiced strong opinions that were taken as factual statements. This was no small matter because these experts were associated with U.S. "national laboratories" at which nuclear weapons are created. Eventually Congress called in the General Accounting Office to render a neutral, third-party judgment.

Read the GAO report *Accuracy of Statements concerning DOE's X-Ray Laser Research Program* and one or more published commentaries on the "flap" such as Deborah Blum's "Weird Science: Livermore's X-ray Laser Flap." Another important exploration of this matter is William J. Broad's *Teller's War.* The GAO report, though not long, might be challenging due to its flat tone and scrupulous disinterestedness. But this attitude is itself a value about which you might want to write. You could contrast the tone and implicit values in the GAO report to the Blum article. Notice that the GAO report was triggered by written objections by Roy Woodruff about how information on X-ray laser research had been misrepresented as "overly optimistic and technically incorrect." Notice, too, what happened to the career of Mr. Woodruff as a result of his whistle-blowing. Write a report showing how values played a role in the representation of factual information and in making opinions appear to be statements of fact.

3. Most of us will never be involved in technical communications of the magnitude of those of the SDI program, with national security, thousands of jobs, and trillions of dollars at stake. Many of us will, however, probably at some time in our careers face ethical dilemmas similar in nature to those discussed here. We might face a question about how to represent technical information in relation to projections and expectations. At what point does a healthy optimism become untenable and unethical? We might face a question about just how accurate and valid is the technical information we are reporting. Though the data might not originate from us, at what point should we raise ethical questions about honesty, for example? Recall the GAO report on the Patriot system that states that invalid estimates of effectiveness could leave soldiers unprotected. At what point does innocently portraying your position favorably go too far, becoming a disservice or danger to your audience? And with whom does the responsibility lie, with you or your employer?

Think of some situations in which you might plausibly find yourself in your career facing these questions. You might want to read the full GAO report on the Gulf War Air Campaign, too. Write a report on how you would answer the questions above.

4. In this chapter, a congressional report on claims about the performance of the Patriot missile system during the Persian Gulf War was discussed briefly. Read this report and write a summary of its purpose, methodology, findings, and conclusions. Include an analysis of the language of the report as to whether the inaccurate representations of information were deliberate or ethical.

What does the report say about the burden of responsibility for verifying the validity and accuracy of technical information? Some people have said the Patriot is like a small-scale SDI system because it intercepts a missile with another missile using a complex computer system. What does the report actually suggest about the likely performance of an SDI system by analogy from the Patriot? Be sure to include details about the proximity for the detonation of Patriot warheads and the validity of indications of "kills" in the Patriot computer system.

There is a series of interrelated documents, any one of which would be suitable for this topic. You could begin at the Congressional Oversight Hearing of the Committee on Government Operations, Subcommittee on Legislation and National Security on April 7, 1991. You could also begin with the GAO Report: *Operation Desert Storm: Project Manager's Assessment of Patriot Missile's Overall Performance Is Not Supported* (GAO/T-NSIAD-92-27, April 7, 1992). You might also want to examine the GAO Report, *Patriot Missile Software Problem* (GAO/IMTEC-92-26). This technical report shows how small software and power supply problems can render the Patriot system totally ineffective.

5. Many of the subsystems of the original SDI program continue to be researched and developed under later, different programs. The current plan for defense against missile attacks has two components: national, strategic defense against long-range missiles (NMD) and defense against shorter-range, tactical missiles within local theaters of operation (TMD). The strategic missile defense has been under development and is moving toward limited deployment (actual operational use) under a plan called "Phase One 3+3." The August 1999 *Scientific American* article "Why National Missile Defense Won't Work," by George Lewis and others, explains that the same problems about SDI continue to plague the latest plan too. Part of this plan is an array ("constellation") of many satellites in continuous orbit around the earth. Some of these satellites, known as Brilliant Eyes, serve as sensors and control stations, detecting missile launches and direction interceptors to destroy them. The other satellites, known as Brilliant Pebbles, are the interceptors or kill vehicles, ever ready to destroy enemy missiles when they reach beyond the atmosphere into space.

Software technology is still a prominent issue even under this narrow subpart of what was to be the overall SDI strategic missile defense plan. Not only are there questions about the effectiveness and survivability of the computers and programs within the Brilliant Pebbles vehicles themselves, but there are also questions about the software simulations that try to demonstrate or evaluate the effectiveness of Brilliant Pebbles in its early development states. The Department of Defense in the late 1980s and early 1990s made several claims about the prospective effectiveness of Brilliant Pebbles derived from computer simulations. Opponents were skeptical of these claims and asked for investigation of them. The General Accounting Office, at the request of Congress, conducted an investigation. Its conclusions are clear from the title of its report: *Estimates of Brilliant Pebbles' Effectiveness Are Based on Many Unproven Assumptions.* Many of these shortcomings in drawing conclusions from computer simulations are almost identical to questions raised about the software meant to run Star Wars, in particular the inability to develop truly realistic simulations of actual attacks. Instead, even the process of simulation has to be refined consecutively through demonstrations of its own inadequacies.

Read this report, write a summary of its analysis, findings, and conclusions, and draw parallels between it and the discussion of SDI in this chapter. Be sure to include comments on the report from the letter by then-SDIO director Henry Cooper, included with the GAO report.

6. In this chapter a sociological report was mentioned that explored the relationship between ethics, technical expertise, and political sentiment in the motivations of those scientists and engineers who signed the Star Wars Boycott Pledge (Nusbaumer et al., "The Boycott of Star Wars by Academic Scientists"). Read this article in full and report on it, emphasizing the interplay of ethics and technical expertise. Summarize, too, the blurring of the distinction between technical motivations and political motivations, as discussed earlier in this chapter. Describe also in brief the gradually increasing politicization of science and technology traced by Nusbaumer et al.

REFERENCES

Adam, John A., and Mark A. Fischetti. "Star Wars: SDI: The Grand Experiment. *IEEE Spectrum* September 1985: 34–35.

Adam, John A., and John Horgan. "Debating the Issues." *IEEE Spectrum* September 1985: 55–64.

Adam, John A., and Paul Wallich. "Mind-Boggling Complexity." *IEEE Spectrum* September 1985: 36–46.

Blum, Deborah. "Weird Science: Livermore's X-ray Laser Flap." *Bulletin of the Atomic Scientists* July-August 1988. 7–13.

Bowman, Robert. *Star Wars: A Defense Insider's Case against the Strategic Defense Initiative.* Los Angeles: Jeremy P. Tarcher, Inc., 1986.

Broad, William J. *Teller's War: The Top-Secret Story behind the Star Wars Deception.* New York: Simon and Schuster, 1992.

Charles, Dan. "The Man Who Promised the Earth." *New Scientist* 20 March 1993: 26–28.

Eastport Study Group. *Eastport Study Group—A Report to the Director, Strategic Defense Initiative Organization.* Marina Del Ray, California: Eastport Study Group, 1985.

Fischetti, Mark A. "Exotic Weaponry." *IEEE Spectrum* September 1985: 47–54.

Fletcher, James. *Report of the Study on Eliminating the Threat Posed by Nuclear Ballistic Missiles, Vol. V, Battle Management, Communications, and Data Processing.* Washington, D.C.: GPO, October 1983.

Haley, P. Edward, and Jack Merritt. *Strategic Defense Initiative: Folly or Future?* Boulder, Colorado: Westview Press, 1986.

Hecht, Jeff. "Blinded by the Light." *New Scientist* 20 March 1993: 29–33.

Johnson, Deborah G. *Ethical Issues in Engineering.* Englewood Cliffs, New Jersey: Prentice Hall, 1991.

Keyworth, George A., II. "The Case for Strategic Defense: An Option for a World Disarmed." *Issues in Science and Technology* Fall 1984: 30–44.

Kogut, John, and Michael Weissman. "Taking the Pledge against Star Wars." *Bulletin of the Atomic Scientists* 42/1 (January 1986): 27–30.

Lamb, John. "The Bugs in the Star Wars Programme." *New Scientist* 21 November 1985: 27–31.

Lewis, George A., Theodore A. Postol, and John Pike. "Why National Missile Defense Won't Work." *Scientific American* 281/4 (August 1999): 36–41.

Lin, Herbert. "The Development of Software for Ballistic-Missile Defense." *Scientific American* 253/6 (Dec. 1985): 46–53.

Mann, Paul. "Historic Turn Eyed in Missile Defense." *Aviation Week & Space Technology* July 5, 1999, 30–31.

Marsh, Peter. "The Anatomy of Star Wars." *New Scientist* 14 November 1985: 31–35.

Nusbaumer, Michael R., Judith A. DiLorio, and Robert D. Baller. "The Boycott of 'Star Wars' by Academic Scientists: The Relative Roles of Political and Technical Judgment." *Social Science Journal* 31/4 (1994): 375–89.

Office of the President of the United States. *The President's Strategic Defense Initiative.* Washington, D.C.: January 1985.

O'Neill, Bill. "Fear and Laughter in the Kremlin." *New Scientist* 20 March 1993: 34–37.

Parnas, David Lorge. "Software Aspects of Strategic Defense Systems." *American Scientist* 73 (September-October 1985): 432–40. Reprinted in *Communications of the ACM* 28/12 (December 1985): 1326–35.

Patel, C. Kumar, and Nicolaas Bloembergen. "Strategic Defense and Directed-Energy Weapons." *Scientific American* 257/3 (Sept. 1987): 39–45.

Pressler, Larry (Senator). *Star Wars: The Strategic Defense Initiative Debates in Congress.* New York: Praeger, 1986.

Reagan, Ronald. "Address to the Nation on National Security." National TV Broadcast. March 23, 1983.

"Rough Sailing for Smart Ships." *Scientific American* 279/5 (November 1998): 46.

Smith, R. Jeffrey. "New Doubts about Star Wars Feasibility." *Science* 229: 367–68.

"Star Wars of the Seas." *Scientific American* 259/3 (September 1988): 14–15.

Steinberg, Gerald J., ed. *Lost in Space: The Domestic Politics of the Strategic Defense Initiative.* Lexington, Massachusetts: Lexington Books (D. C. Heath and Co.), 1988.

Strategic Defense Initiative: Progress and Promise. Department of Defense. Washington, D.C.: U.S. GPO, 1989.

Tirman, John, ed. *Empty Promise: The Growing Case against Star Wars.* The Union of Concerned Scientists. Boston, Massachusetts: Beacon, 1986.

———. ed. *The Fallacy of Star Wars.* The Union of Concerned Scientists. Boston, Massachusetts: Beacon Press, 1986.

U.S. Congress. Office of Technology Assessment. *Directed Energy Missile Defense in Space.* OTA-BP-ISC-26. Washington, D.C.: GPO, April 1984.

———. *Ballistic Missile Defense Technologies.* Princeton, New Jersey: Princeton University Press, 1986.

———. *SDI: Technology, Survivability, and Software.* OTA-ISC-353. Washington, D.C.: GPO, May 1988.

U.S. General Accounting Office. *Accuracy of Statements Concerning DOE's X-Ray Laser Research Program (Strategic Defense Initiative Program).* GAO/NSIAD-88-181BR. Washington, D.C.: June 1988.

———. *Estimates of Brilliant Pebbles' Effectiveness Are Based on Many Unproven Assumptions (Strategic Defense Initiative).* GAO/NSIAD-92-91. Washington, D.C.: March 1992.

———. *Operation Desert Storm: Project Manager's Assessment of Patriot Missile's Overall Performance Is Not Supported.* GAO/T-NSIAD-92-27. Washington, D.C.: April 7, 1992.

Waldrop, M. Mitchell. "Resolving the Star Wars Software Dilemma." *Science* 232: 710–13.

WEB SITES

Computer Professionals for Social Responsibility (includes Risks Digest recording problems in use and claims about computer systems): http://www.cpsr.org/cpsr

Federal of American Scientists, which provides excellent, detailed background material and summaries of key defense developments including missile defense: http://www.fas.org

U.S. Ballistic Missile Defense Organization link for news and information about ballistic missile defense: http://www.acq.osd.mil/bmdo/bmdolink/html/bmdolink.html

U.S. Department of Defense, "Defense Link," includes news articles, photos, and reference sources: http://www.defenselink.mil

CHAPTER 8

ETHICS EXERCISES

In this chapter you will find six exercises on ethics in technical communication. They are in the form of hypothetical cases that present a fictional but realistic technical communication situation. Imagine that you are the technical communicator faced with the ethical dilemma described. Each case will describe the circumstances and the relevant factors. These exercise cases are less momentous than the real cases examined in earlier chapters. But in a sense they are also more realistic than the earlier cases because they are more typical of communication situations that the average technical communicator might actually encounter.

Read each case carefully and reflect on how you would handle the situation if you were the actual technical communicator involved. Come to some definite decision as to what course of action you would take, and write down in coherent paragraphs how and why you came to that decision. This written explanation need not be logically rational like a lawyer arguing a case. But you should *try* to be as clear as you can about why you decided as you did in a statement that tries to communicate meaningfully in terms others can understand and relate to.

The "try" is emphasized here for two reasons. First, our efforts to try to communicate in terms meaningful and understandable to others do not necessarily guarantee that they will be received that way. The point, though, is the trying. Second, the exercise asks you to articulate a rationale for your ethical judgment. This can be done only after some reflection not only on what your final judgment is but also how and why you arrived at it. This reflection will help you to understand, perhaps more clearly than before doing these exercises, how complex realistic ethical situations can be. It will also help you to understand yourself and your ethical values perhaps more clearly and with greater confidence than before. Again, the point is not so much the results but the trying.

These exercises can be done individually by yourself, collaboratively as a small group undertaking, or in a combination of the two methods as you revise an individual statement after group discussion with and insights from others.

Note to Instructors: As an aid to instructors, a plausible ethical analysis is offered for each case. This brief analysis is intended only for exploring many of the possible angles of a case, not for rendering a definitive judgment. The purpose of

233

these comments is only to suggest possible ethical dimensions of the case that you might not have uncovered on your own. These dimensions can then be used to facilitate and stimulate class or small-group discussions. The sample analyses are presented together at the end of this chapter.

In addition to having students write and discuss their own responses, you could also ask that they report on how each of the ethical perspectives we have discussed would apply to the case. The intent of this extra task is not to suggest that abstract theories are more important than personal judgment, but only to practice seeing ethical situations through many different ethical lenses and to help expand our horizons of ethical awareness.

CASE ONE: LASER GLITCHES

You are a technical communicator for Janus Industries, a major defense contractor. Janus has been developing a new laser targeting system for large guns on tanks and other armored vehicles, which is now in the final stages of testing. Your firm is very pleased with the system and eager to begin deliveries to the U.S. Army even sooner than the contract calls for.

Extensive usability testing of the weapons system both in simulations and under realistic field conditions has shown that it meets or exceeds all the specifications stipulated in the contract with the army. In a very few instances under highly unusual conditions, however, the system has acted up unexpectedly. Specifically, in high-speed maneuvers when the gunner is very fatigued or highly stressed and during conditions of low visibility in a dusty environment such as a desert, the gunner does not accurately carry out the input commands in the proper format asked for by the system. The system then crashes, becoming inoperative for several minutes before it can be reactivated.

In the six months of field testing, this problem was encountered only three times. Even in those cases, the tank was executing maneuvers that were not explicitly called for by the contract. These maneuvers were only part of your firm's internal testing criteria.

The cause of the problem is a combination of factors including the gunner's fatigue, poor visibility, the bouncing vehicle, the weak brightness of the display panel, and the ambiguous wording of the commands displayed. In any event, the computer software and hardware driving the system both operate entirely correctly. Therefore either the display itself or the interaction between the gunner and the display subsystem seems to be the crux of the problem. That is, the whole system in itself works properly, though the operator at times does not operate it properly.

As a technical communicator, you have been working closely on this project, writing technical documents and progress reports. You have just been asked to write the formal final report on the field testing, the submission of which to the army will precede the first deliveries of the targeting system. Your earlier interim reports taken together will supply nearly all the information you will need for this final report. Now you are just pulling together the material from your earlier reports and putting a finishing touch on the package.

That at least is what your supervisor expects you to do. Both she and you are aware of the very infrequent glitches with the system, however. She expects that these exceptional instances will not appear in your final report because their circumstances were not called for by the army.

You feel uneasy, however, because your report will not be reflecting all the relevant information that your firm has learned about the operating characteristics of the system. If you do what your supervisor expects, you would not exactly be lying, though neither would you be entirely honest. You are doubly concerned because of the potential serious harm that could result from this problem, rare though it may be. Not only would the possibility of attacking an enemy tank be lost temporarily, but your own tank might be put in jeopardy, too. You recall the old military saying, "For want of a nail, a shoe was lost." You decide to talk to your supervisor about your concern.

She tells you directly not to report the infrequent problem. After all, she says, this happens only rarely under highly unusual conditions. The likelihood of its occurring in the actual use of the system anytime soon is negligible. Besides, the country is at peace, so no harm would be done to anyone should the problem ever crop up in training exercises. In the meantime, she says, she will bring the problem to the attention of higher management and engineers so that they can correct the problem as soon as possible. Once they correct the problem, they will report it to the army as an update on the system. Thus in the long run the truth will be told, and the problem will be corrected. She asks for your report by the end of the week. What do you do? And why? And how is this ethical?

DISCUSSION

You have two basic options, of course: to report or not to report. If you were to leave out the information, you might not feel you are doing anything unethical. After all, you are just giving the army precisely the information they specifically asked for. In addition, you are doing what your employer and your supervisor want you to do and have specifically directed you to do. On top of that, you have informed your supervisor of your concern, who supposedly will inform upper management, who in turn will initiate corrective actions. It would seem that you have done what is specifically expected of you, and that should be enough.

Another important issue for you personally is the security of your livelihood. You are the sole means of support for your family of four. The national economy at this time is not strong, and the defense industry has been hit hard by the recent government's cutbacks of defense spending. You are lucky to have your job and want to do everything you can to keep your job; after all, your family is counting on you.

The other option is to include the uncalled-for but relevant information in your report. Thus you would be informing the army of something they really should know about. On the other hand, once your supervisor sees your report, she might insist that it be revised to leave out that information. Should that happen, you would leave out the information, as ordered. You could just leave it at that, with the ethical satisfaction of having taken some positive action. Or you could contact the army separately, either conspicuously as a whistle-blower or perhaps as an anonymous informant.

Variation

After you have thought through the exercise and written your explanation of your ethical decision, consider this variation on the case.

The United States has just entered a small but vicious local war on behalf of an ally. In this war the Janus targeting system will be put to its first combat use—and very soon. This targeting system is especially well suited to these particular battlefield conditions. Some of the battlefield consists of desert, so you know the glitches will likely arise soon.

Would you act differently in this variation? How? Why? If you would decide to act differently, what does this say about the values underlying your decision? Are you a person who is flexible in your ethical decision making, taking all the factors into consideration? Or are you a person whose sense of ethics insists instead on the importance of fundamental principles rather than shifting circumstances? Consequentialist theories such as utilitarianism emphasize the effects that follow from your ethical decisions. Another major class of theories, which includes Kant's theory, emphasizes duty above all else. How relatively important is the security and comfort of you and your family compared to a remote possibility to a nameless tank crew?

CASE TWO: FREEDOM

You are a Web consultant who has been hired by an individual to post the details of a complex chemical process to the Web. As you work, you notice details about the toxicity of chemicals, the concentration levels of aerosols, and some references to military tests of various means of delivery. You have been told that you are only putting on the Web some factual information that is already available to the public through other means. It becomes clear to you, even with your limited knowledge of chemistry and biology, that these documents explain how to make chemical warfare agents and delivery devices and how to deliver them surreptitiously for maximum effect. You are actually helping to write a terrorist technical manual.

When you ask about the purpose of posting this information, you are told to mind your own business. Your client also tells you that how this information is used is neither his business nor yours. After all, he says, all you are doing is posting simple factual information. How that information might be used is the responsibility of the user, he says. The information is like what is said about selling handguns: Guns don't kill people; people do.

You have not had much business lately due to a downturn in the local economy. You need the business, and your client has alluded to the possibility of steady work for you for a year or more. What do you do?

Variation

If you originally chose to take action to notify some legal authority about this, you apparently feel that your responsibility for the potential dangers involved outweighs your client's freedom rights. You might also have refused to continue this assignment. Can you imagine a similar situation but with information of somewhat less dangerous potential in which you would *not* reveal this activity to authorities? Use this variability to clarify the threshold for what you consider ethically compelling.

If, on the other hand, you originally chose *not* to take any action to reveal this to authorities, you apparently feel that the rights of the client to free expression are more important than the harm that this information might lead to by someone else's hand. Can you imagine a similar situation but with information even more dangerous or repugnant to you, in which you *would* reveal this activity to authorities? Use this variability to clarify the threshold for what you consider ethically compelling.

Discussion

In an earlier chapter on information from the Nazis, we saw how the way information was obtained and how it might be put to use in the future have a great deal of ethical significance. Information does not exist in a vacuum but is almost always sought for some purpose, which may be or may not be innocent.

Consider the cruise missile attack by the U.S. military on Osama bin Laden in 1998. One of the targets, a factory building in Sudan, was said to have been used to manufacture chemical warfare agents including the VX nerve agent. The evidence for this claim was that the presence of an exotic chemical precursor of VX had been found in the soil near the plant. Many years ago, when the United States had an active chemical warfare program, VX was one of its most potent discoveries. The information on how to produce this deadly chemical warfare agent, ironically, was released unwittingly by the United States during a broad movement to declassify previously classified information and documents. Thus, ironically, terrorists can now utilize this information against the United States or other nations. The inadvertent communication of this technical information has put at risk potentially thousands of people. It is also common knowledge that detailed information on how to construct or manufacture explosives is now freely available on the Web to anyone, including terrorists.

CASE THREE: CONFIDENTIALITY

You are employed as a technical communicator by Caduceus Company, a major provider of software systems for the health care industry. Currently you are working on help-desk instructions for a new software system that operates from a database accumulated from all the records of all the hospitals, clinics, HMOs, laboratories, and physicians in your state. These records carry a good deal of highly personal information on patients and are, of course, private and confidential. The revelation of some of this information to the wrong people could be seriously damaging for those patients. Improper release of this information could, for example, limit their employability, damage their reputations, or restrict their access to health insurance at affordable rates. Part of your job, in fact, is to ensure that confidentiality is maintained by all who use the system while preventing access to those who might misuse it.

Just yesterday you learned of some shocking information. While observing the typical daily operation of the help-desk staff in resolving users' questions, you recognized the name of one of the clients called up on a screen. It is the name of your cousin's fiancé, and the entry indicates that he has tested positive for HIV and has received counseling about HIV and AIDS from a local clinic. You and your cousin are very close, and you are certain that she is unaware that her intended spouse has tested positive for HIV. The marriage is only a few weeks away. She would be highly endangered after marriage but also in their current relations without this knowledge.

Do you reveal this information to your cousin but violate the privacy and confidentiality that you have sworn yourself to and are legally obliged to maintain? Or do you keep the information to yourself but jeopardize the health and life of your cousin by preserving the deception by her fiancé?

Variation

Let's change the circumstances a bit. Suppose, alternatively, that the information was only about a neighbor on your block or a distant cousin of your neighbor? Would this make a difference in your decision to disclose or not? What does the change or the lack of change in your decision say about the values at work in your decision? Does personal closeness play a great role in your ethical decision making? But everyone is close to someone else, so should not everyone be granted the same importance in anyone else's ethical decision making? Or is the strength of a general, universal principle more important than the particular details of a particular personal situation?

Let's make another change. Suppose that what you learned about was only a relatively benign genetically transmitted disorder that expresses itself in only a small percentage of those carrying the genetic marker. Even then it appears only in the form of occasional minor muscle spasms that might limit some physical activities but could not be considered a serious disability or impairment by any stretch of the imagination. In this alternative situation, the threat is not to your cousin but only to her possible offspring (none of which are planned) but only at a low level of probability and only to a mild degree. Again, would your response be different from what it was in the original case? And what does your answer now in comparison to your earlier answer imply about the values at work in your ethical decision-making process?

CASE FOUR: BURIED INFORMATION

Suppose that you are on the technical writing staff of Isomer, Inc., a chemical manufacturing company. In your position, you have access to and must be familiar with the documents that form part of the institutional history of your company. One day, as you are researching the archives for your current project, you encounter a disturbing series of documents.

These technical and scientific documents relate to research on the health and environmental hazards of the chemicals your company produces. The archives reveal that the original drafts of these reports contained clear, firm conclusions that some of these chemicals are very hazardous under long exposure to them. The final, revised drafts, however, showed only watered-down, tentative conclusions.

You also find reports of unusual illnesses among workers exposed to certain chemicals. The reports do explore a possible link between the illnesses and exposure to the chemicals but, to your mind, they seem to be straining to explain away the findings as coincidental or inconclusive of any serious hazard. Upon looking further into the archives, you find that some of these reports are so serious in nature that by law they should have been communicated to the Environmental Protection Agency (EPA) and to the Occupational Safety and Health Administration (OSHA).

What do you do with your new knowledge? Do you leave things the way you found them and not report the information to EPA and OSHA? Or do you bring this health hazard to the attention of your supervisor? Or do you assume that your supervisor and higher management will squelch the information, and so you should leak the information to EPA and OSHA on your own because of its serious nature?

CASE FIVE: WHISTLE-BLOWING

This case is an extension of the preceding Case Four. Assume all the same content as in Case Four: Buried Information, but now consider it in a different light. Set aside for the moment the various ethical theories we have been discussing. Focus instead only on the question of what are the conditions under which you should blow the whistle, and what are our ethical responsibilities in whistle-blowing generally.

In the chapter on the shuttle *Challenger* disaster, we learned that Roger Boisjoly acted very responsibly in blowing the whistle on the actions of the management of Morton Thiokol and in criticizing NASA management both before the disaster and afterward. We also learned of the misery this brought down on him. At first he was threatened with being relieved of his job. In response, Boisjoly threatened to sue his employer under the federal whistle-blowing laws, which would have involved a great deal of time, money, and emotional energy. Though he won back his regular position, he still found his assigned tasks and working relationships so difficult and unpleasant that he voluntarily resigned after a while. The costs to him were high.[1]

Part 1.   This case has three parts that can be done in any order or separately. In the first part of this assignment, read several articles. Mike M. Martin's "Whistle-blowing: Professionalism, Personal Life, and Shared Responsibility for Safety in Engineering" springs from an engineering context, yet its exploration of the issues can apply equally well to practically all professions, including technical communication. Martin deals with the connection between our professional and personal lives; the responsibility of the public to support and protect whistle-blowers; and the different senses of the term "virtue" that might apply in such cases. Write a brief report on this article, and explain how it might apply to any of the cases we have considered.

Joseph H. Wujek's "Must Engineers Behave Heroically?" also springs from an engineering context yet can apply very validly to technical communication, too. Wujek argues that engineers are *not* ethically required to behave heroically, sacrificing themselves and their careers for the sake of abstract principles alone. All that is ethically required is that we behave as other reasonable, typical people would do in the same situation. Apply Wujek's perspective to this case. Would you be comfortable as a member of society or as a worker in this plant if those with special knowledge held an ethical perspective that did not call for potential sacrifices by them? Would you, on the other hand, be comfortable as the actual discoverer of this information with an ethical burden potentially calling for self-sacrifice such as losing your job and being blackballed?

Mike Markel's "A Basic Unit on Ethics for Technical Communicators" offers a broad view of the network of ethical ties we have to those around us. We do, of course, have obligations to our employer, but likewise our employer has definite obligations to us, too, as well as to the public. What are some of the ethical responsibilities discussed by Markel that we have not already discussed in this

---

[1]For a brief summary of Boisjoly's experience at Morton Thiokol, see "Engineering Ethics Cases" at the ethics Web site maintained by Case Western Reserve University, http://www.cwru.edu/affil/wwwethics/engcases.html. This site also includes several other notable real cases presented in detail, eleven other real cases through Texas A&M, a section on applied ethics cases of the month, and many hypothetical cases to stimulate discussion. It also includes as one of its detailed real cases that of architectural engineer William LeMessurier discussed later in this and another hypothetical case.

book? Discuss how these additional ethical dimensions might operate in your own professional life in the career position you are now in or are working toward.

Keep in mind that deciding to blow the whistle is usually founded on several assumptions. It assumes that you have the big picture and that all the most important, relevant information and issues are known to you. This assumption might not, however, reflect reality. In this case, for instance, there might exist additional tests or research that had been conducted that disconfirms any connection but which you have not become aware of yet. It assumes, too, that your judgment is more correct, more true, more reasonable than that of the writers and managers. To some extent, the firm does have a right to draw conclusions that it genuinely feels are warranted in light of the facts and their own experienced judgment. The key question, of course, is, What is the extent of that right? Recall one of the statements from the *Challenger* investigations, in which a supervisor was told to "take off your engineer's hat and put on your manager's hat."

Part 2. In the second part of this assignment, assume, on the other hand, that sometimes you just have to work with what you have and use it as best you can. The federal government has instituted whistle-blowing laws over recent years and provided institutional means to support whistle-blowing where appropriate. A good example of this can be found at the Office of Research Integrity. Though this office deals with scientific research (for the Department of Health and Human Services, Public Health Service), its explanation of the principles involved in determining when whistle-blowing is warranted is applicable to many fields, including technical communication.

Read the DHHS, PHS Position Paper #1, *The Whistleblower's Conditional Privilege to Report Allegations of Scientific Misconduct.* Summarize and report on its key points, especially its description of when whistle-blowing would not be appropriate; its clarification of how much information is sufficient cause for whistle-blowing; and the institutional means for supporting and defending whistle-blowers against retaliation. Also read and report on Appendix A, *Responsible Whistleblowing: A Whistleblower's Bill of Rights.* Both documents are available by calling or writing ORI or at their Web site (http://ori.dhhs.gov). You might also want to research other governmental, legal, academic, or professional sources on whistle-blowing.

Part 3. In the third part of this assignment, consider the question, When do you blow the whistle on yourself? The case of the architectural engineer William J. LeMessurier involves blowing the whistle on a flawed structure that he himself designed, the new (in 1978) 59-story Citicorp corporate headquarters in New York City. A number of sources are available describing this case, but the most accessible is probably Joe Morgenstern's article "The Fifty-Nine-Story Crisis." This real-life story is not only good ethics but also high drama.

Read Morgenstern's account (or any other detailed account) of this episode (see also the Web site in footnote 1 regarding Boisjoly). Report on LeMessurier's dilemma, how it was brought to his awareness, how he decided on it, what was done about it, and the ultimate effect of whistle-blowing on his career.

CASE SIX: GOVERNMENT ETHICS

You are a contract technical communicator currently working for the Department of Health and Human Services. Your job is to produce a series of pamphlets for the public as part of the department's outreach activities. This job has two sides to it. One side involves transforming a mass of statistical data from many sources into a series of meaningful informational messages. The other side involves translating the department's regulations and policies, written in bureaucratic style, into a series of pamphlets aimed at the lay public audience to inform them of the department's programs, procedures, and policies.

The informational task is somewhat of a problem. You believe you are expected to cast the information in a light that does not seem justified by the data as a whole. Some of your drafts are returned with notations to use other data. Sometimes you are told to use older data rather than newer data, sometimes from one study rather than others, for no apparent reason. It seems to you that the incidence of poverty, childhood diseases, and malnutrition for different areas of your region is being selectively adjusted and misrepresented.

The translation task is also a problem. It seems to you that word choice and style are being guided by traditional assumptions that are no longer valid. This stylistic preference strikes you as unethical because it suggests that the department is affirming societal assumptions that are exclusionary and foster discrimination. Your drafts are often returned using the male pronoun "he" for a generic person in the pamphlet examples. Examples of families all include married couples with both partners present, though the reality of most of your audience is very different. The examples that you drafted that mention the needs of gays or lesbians are dropped outright without explanation. You are concerned that the actual needs of your audience are not going to be served by the selective information in these revised pamphlets. You are also concerned about the rights of others simply to be recognized as an integral part the fabric of our nation. What do you do?

Discussion

The federal government has many ethical guidelines in place and usually has a specific office within each agency to handle only ethics issues. It would be useful to become familiar with some of these guidelines. For this case, you need to consult the ethical guidelines of your particular department or agency. You could also consult the government agency overseeing the ethics policies of all the government's other departments, the Department of Justice.

These governmental guidelines represent an approach to ethics that is rather different from the approach used in this book. Notice that the primary office responsible for ethics policy is the Department of Justice, which makes ethics a matter of explicit and specific *law* rather than a matter of personal interpretation and responsibility. Other departments may have their own guidelines elaborating the Department of Justice rules into a form appropriate for their activities.

Notice too that these guidelines are formed into specific categories of activities. These categories include conflict of interest, confidentiality, impartiality in carrying out policies, financial disclosure, outside activities, and other categories such as the appearance of impropriety. This has the advantage of making ethics as explicit and specific as possible with as little latitude as possible for selective, self-serving interpretations. A fixed number of categories of ethics such as this might have limitations, however. It might exclude other activities that do not fall into these categories but that nevertheless seem to be unethical, but they leave the suggestion that they are not really problematic. Keep in mind that these ethical guidelines have a very specific context and purpose, namely the administration of the federal government involving accountability for the use of public tax dollars in a context of strong rights for equal individuals.

Read and summarize the Department of Justice ethical guidelines. Do the same for one other federal department or agency that you might be dealing with some day as a technical communicator. Many of the Web sites for the ethics programs include games that provide training on ethics. Some of the games can be played online or downloaded to be played later. Here are some examples:

- Department of Justice's game *QUANDRIES:* (http://www.usdoj.gov/jmd/ethics/)
- United States Office of Government Ethics's *Ethics Training Games (need to be downloaded):* (http://www.usoge.gov/usoge006.html#games)
- Defense Information Systems Agency's (part of Department of Defense) *Ethics Interactive Training:* (http://www.disa.mil/ethics/html/)
- Department of the Interior's *Automated Ethics Training Program:* (http://www.doi.gov/ethics/ethics.html)

Some other departments have access open to the public, whereas others have access restricted to members of that department.

Are the ethical theories that we have been studying represented in the guidelines you have chosen to explore? If these theories do fit these guidelines, describe how. If they do not fit, describe how they do not.

APPRAISAL, CASE ONE: LASER GLITCHES

Aristotle. From an Aristotelian ethical perspective, you should take action to do what is good, that is, protecting those who might be endangered if you added nothing to your report. These people, the stakeholders including the soldiers but also your country, trust you and depend on you and your firm to present them with honest and complete information and reliable equipment. To do nothing would be dishonorable because it would violate their trust in you and your firm.

To be sure, Aristotle is known as both a theorist and a pragmatist. The pragmatic side would urge that you recognize that the responsibilities involved here are not yours alone. You could begin to discuss this with your supervisor as a question that has just occurred to you rather than as an ultimatum. You could also ask for guidance from your supervisor about how such issues have been dealt with in the past or what company rules might offer guidance in such matters. You could take a pragmatic approach by offering a suggestion that accomplishes both the needs of the client and the wishes of your firm. This could be done by adding a brief but obvious caution box explaining what might happen if certain very rare circumstances should arise. You could also caution to avoid such circumstances as much as possible and provide instructions on how best to handle the glitch if it does occur. Or you could include in your report some brief acknowledgment of the problem by your firm and offer a plan and a time for rectifying the problem as soon as possible. Your report could also affirm that your product meets all the criteria stipulated in your contract. In drawing this problem to the attention of the customer, you are doing them a service by performing above required expectations. This would put as positive a cast as possible on what is otherwise a shortcoming.

Kant.   From a Kantian perspective, the consequences to yourself are much less important than your duty to others and your country. Even though you might jeopardize your job, you should include the information in your report. If your supervisor has the information removed, then you should communicate that information clearly and effectively to the army through another means. Remember that treating others as you would wish to be treated is central to Kant's perspective. No one would want to be the poor tank gunner who is left without a usable gun in the heat of battle.

Utilitarian.   From a utilitarian perspective, the case is somewhat complicated. In peacetime, there would be only negligible consequences if you failed to report the whole truth: infrequent episodes of problems, with little significance, that would be appearing in peacetime and then only during the short period during which the problem is being solved. The potential cost to you and your family would be very great, however, if you were to blow the whistle. Therefore you probably should chose not to report the whole truth.

In wartime, however, the situation would be substantially different. The stakes would be potentially very great not only for the gunner and his crew but also the whole military group (Recall, "For want of a nail, . . ."). Most people would agree that in wartime the calculation of cost versus benefits would dictate that you jeopardize your job for the sake of potential greater benefits to many others.

Feminist Perspective and Ethic of Care.   From a feminist ethical perspective, the case is again somewhat complicated. Granted, most feminist thinkers would deplore being involved with the military in the first place. Carol Gilligan and other feminist ethicists have pointed out a fundamental difference between men and women as to how they deal with ethical dilemmas. Gilligan says that men typically insist on fundamental principles of justice. These principles are universal and must be applied inflexibly regardless of circumstances. Gilligan herself (who does not stand for all or most feminists) also says that women typically insist on the primacy of relationships and feelings for others out of a caring concern for them, which requires that all the circumstances and above all the actual people involved should determine one's ethical judgments.

The basic caring concern found in ethics of care such as Nel Noddings's stresses an excellent basis for approaching a dilemma but does not necessarily make the rendering of a decision straightforward. Caring concern for whom and to what comparative degree? A caring concern for our family emphasizes the strong likelihood that you would lose your job if you reported the whole truth. You and your family then might face impoverishment, stress in many forms, and the disruption of having to find another job perhaps in another state. Should this outweigh a caring concern for a few unknown soldiers in a distant land who face a remote possibility of being harmed, though the harm to them would be profound if it were to occur?

NB: This is a genuinely realistic case. The U.S. Army's M1A1 main battle tank suffered some perplexing problems with its highly sophisticated gun-targeting system during tests as well as with other electronics. Among the problems was "uncommanded rotation of the turret," which occurred infrequently but of course should never occur at all. Under actual battlefield conditions, this glitch could have life-or-death consequences. Whether and how this was to be reported was a matter of debate among those tasked with reporting on the tests.

APPRAISAL, CASE TWO: FREEDOM

Aristotle. From an Aristotelian perspective, a refusal to continue to post the information and take steps to inform appropriate authorities would probably be in order. It is the good thing to do in the sense of protecting innocent lives and the right thing to do in the sense of putting the civic good above your own job or your client's motivations. Athenian politics and philosophy were very strong on the vital centrality of the society over and above any other interests. Historians tell us that even our modern notion of "the individual" existing apart from the culture and society was unknown in Aristotle's world because the self was the *civic* self. The only potential complicating factor in this case would be that the law was also fundamentally important to Aristotle and his world. If the law were entirely clear and absolute on granting unlimited freedom to publish anything at all, then from an Aristotelian perspective emphasizing abiding by the law, you would have to continue with our posting. At the same time, however, the courts would offer a forum for the full interpretation and implementation of the law to be publicly debated.

Kant.   From a Kantian perspective, your duty is clear: Stop the task and take action to prevent innocent people from being harmed. This is the way practically anyone would want to be treated if they were the ones potentially at risk from these chemical weapons.

This obligation is binding, so the impact on your income would be largely irrelevant. It is true, however, that some contemporary thinkers feel that the impact to yourself *should* be a factor in determining one's duty because we have a duty to look out for ourselves, too. This is not the traditional Kantian perspective, however.

Utilitarianism.   From a utilitarian perspective, the ethical thing to do would be to discontinue your work and inform the authorities. The benefit to your client in exercising his or her freedom would be vastly outweighed by the potentially great costs to untold numbers of others. If you chose to discontinue the work, the cost to you of losing your job would be miniscule compared to the great benefit to others. Adherence to abstract principles such as freedom, as this example shows, is not a factor in the weighing of costs and benefits for most utilitarians.

Feminist Perspective and Ethic of Care.   From a feminist perspective, in general one does not have a right to harm others or to foster such harm. There is no such thing as ethically neutral or indifferent communication because all communication and language use inherently entails values. Therefore, the posting should stop. From an ethic of care perspective, caring and positive relationships are paramount of course, so discontinuing the work and informing the authorities would clearly be the proper decision in this case. Our concern for the harm that might be done to so many others should override our concern to protect the rights of our client to free speech. Our concern to have a trusting, constructive relationship with the public would also override free speech. The same is true for the welfare of our society as a collective relationship, the continued existence of which might be violently and unfairly threatened by terrorists.

NB: An interesting parallel to this case deals with how to build an H-bomb. In 1979 a graduate student, Howard Morland, tried to determine whether he could find out how to build an H-bomb using only information available to the public (that is, unclassified information) at that time. He succeeded, building on information gleaned from many different sources. Remarkably, the most crucial item of information was obtained from an essay by Dr. Edward Teller in the *Encyclopedia Americana!*

Morland went public with this knowledge, publishing an article on how to build an H-bomb in *The Progressive.* Apparently he got it right, for the Department of Defense immediately obtained a court order to silence him, to cease his publication, to seize the journal's plates, and to confiscate all the issues already published. Morland went to the U.S. Supreme Court to show that all this information was already public and so he was not releasing any classified information. He won his case and was allowed to resume publication.

For a concise summary of these events and a brief interview with Morland, together with very striking visual aids, see Robert Del Tredici's *At Work in the Fields of the Bomb.*

APPRAISAL, CASE THREE: CONFIDENTIALITY

Aristotle.   From an Aristotelian perspective, a case can be made either way based on important virtues. On the one hand, fairness and our natural caring concern for a relative would lead us to decide to disclose the information. It is the good thing to do, certainly from the point of view of your cousin. It is also the true thing to do, revealing the truth while refuting an untruth. Whether it is right or not is less clear, however.

On the other hand, on the basis of civic values such as justice and the need to uphold the laws that define our country, one would have to decide not to disclose the information. To decide this way has the additional merit of affirming the social goal of maintaining good order through the rule of law, which no one is above.

Our judgment should be tempered by wisdom and mercy, Aristotle could well say, which should guide us in deciding when exceptions may be made to our usual practices. This approach would also reflect the Aristotelian avoidance of extremes, in this case avoiding the extreme of applying the law in a totally inflexible way. Aristotle would likely have chosen to disclose due to the basic injustice, mistrust, and unkindness shown by the fiancé. These vices (the opposite of virtues) would warrant our suspending ordinary rules for someone who flouts the rules.

Kant.   From a Kantian perspective, the ethical course of action in the original situation is clear (though illegal, we should note). We should act as we would expect others to act toward us and in ways that should have universal applicability. If we were the cousin, all of us in all honesty would likely much prefer to learn about this secret yet vitally important information. Our very life is at stake, after all, as well as our long-term happiness in several ways. In addition, though, can we reasonably wish that the principle behind this course of action should have binding universal force? Again, most people probably would agree that the seriousness of the stakes so heavily tip the ethical scales against the side of blind compliance with the law that the ethical judgment to disclose should be binding on all people in such situations.

The universal applicability of the Kantian perspective, indeed, is the crux in the first variation of this case. Now the closeness of our relationship to the unaware partner is weakened, though the seriousness of the potential consequences to the unaware partner remains the same. From the point of view of Kant, if we chose to disclose in the original case, we would have to choose to disclose in the first alternative, too, because the particular persons involved should be irrelevant. The abstract and universal principle is everything for Kant.

In the second variation, though, the seriousness of the consequences is weakened considerably. In this situation, from a Kantian perspective most people would, I think, consider the seriousness of the potential consequences to be not so great as to warrant a clear violation of the law. We should, then, follow the ethical principle that valid laws should be honored (unless extraordinary circumstances clearly warrant otherwise). We should note here that the application of an abstraction such as Kant's principles of duty can sometimes take a form rather similar to utilitarianism.

Utilitarianism.   From a utilitarian perspective, the calculation of cost versus benefits is complicated. We need to calculate the tangible costs (such as losing your job) and benefits. We also need to calculate the costs and benefits of intangibles such as pain, suffering, violated trust, and a broken relationship. Most people would agree that the potential benefits are so exceedingly great compared to potential costs, including to yourself, that a decision to disclose the information would be ethically necessitated.

In the first alternative situation, the costs and benefits in an absolute sense would remain the same though the relationship to you is different. The utilitarian perspective is supposedly indifferent to persons and relationships. This indifference is usually considered to be a strength in that it puts everyone on an equal footing, neglecting irrelevancies such as rank, status, wealth, race, gender, or personal relationship. The utilitarian maxim is simple: The greatest good to the greatest number.

In practice, of course, things just are not that simple. Though our democratic notions of equality in principle would put the president of the United States on the same footing as anyone else, the pragmatic reality is that that person, whoever he or she might be, has abilities, knowledge, and responsibilities that few other people have. There are many conceivable situations, such as national emergencies or war, in which the greater good for the whole nation would be served by preserving the health and life of the president rather than someone else. Thus the good of preserving the president would accrue not only to that particular individual but to very many others as well. The same might be said of a medical doctor during a national emergency. The bottom line for a utilitarian appraisal in this situation is difficult to determine.

In the second variation, relative clarity returns. The potential benefits of disclosing in terms of the avoidance of pain and suffering for the offspring, coupled with only a weak likelihood of any problems at all, are comparatively low even in the worst-case scenario (though we would wish this fate to befall no one, of course). One benefit of disclosing would be a strengthened bond of trust between you and your cousin. On the other hand, the potential costs are comparatively high. You might lose your job, negatively affecting not only you but your family, too. Your future employability might also be jeopardized. In addition, you have deliberately chosen to break the law without a clearly compelling reason, for which you could be legally punished.

Others might be affected, too. The reputation of your employer might be damaged, negatively affecting the livelihoods of many other employees in the company. In addition, other people who might suspect that they are infected with HIV might put off being tested and so miss out on timely, effective treatment while jeopardizing the health and lives of their partners. Cost-benefit analyses are hard to compute partly because of this rippling of effects. The bottom line would therefore be not to disclose the information.

Feminist Perspective and Ethics of Care.   From a feminist perspective, in the original case your decision should be fairly clear. Though some feminist ethicists might disagree, most would emphasize the need to affirm the right to health and happiness of anyone over the right of an impersonal organization to keep important information to itself. Naturally, an ethics of care position would emphasize the caring concern for the health—even life—of your cousin over the inflexible adherence to an abstract principle of justice regardless of the particular situation. Though your cousin's fiancé does have a right to privacy, the very high potential for grievous harm to your cousin if she were to remain unaware of this information is of much greater importance. Many feminist thinkers assert the central importance of relationships, too. The fact that your cousin's trust in her partner is unwarranted and that the relationship is not as honest as she might think it is, should then be important factors in judging this case.

In the first variation, it is not completely clear how to decide. It could go either way from a feminist ethical perspective. In the second variation, though, considering in particular the ethics of care of Carol Gilligan and Nel Noddings, adherence to the law and to the ethical principle of privacy and respect for the wishes of others would seem to take precedence over an unlikely and mild threat. Thus we see that the severity and likelihood of the consequences can play an important role in ethical decision making.

NB: Should you wish to pursue this topic in detail, Reid Cushman's "Information and Medical Ethics: Protecting Patient Privacy" is an excellent resource. Cushman identifies most of the important areas of concern and offers a clear analysis of the forces driving us toward increased dependence on electronic records in our society and of the privacy concerns this raises. Especially in the absence of clear, comprehensive, and strong laws ensuring the privacy of medical information, the ethical burden to protect privacy is doubly important as a bulwark against abuses.

Another avenue you might pursue, along the lines of rhetoric linked to ethics, is to examine the values that are operative in the generation and dissemination of medical reports. Though medical descriptions often appear to be completely objective, neutral, and disengaged from value issues, an article in the *Hastings Center Report* (a journal dealing solely with medical ethics) explains otherwise. Suzanne Poirier and Daniel J. Brauner in "Ethics and the Daily Language of Medical Discourse" rely on Jacques Derrida's ethical criticism of language use to explain how medical language is heavily loaded with value implications even though it appears not to be. Using the medical reports presented by medical students on their daily rounds, Poirier and Brauner show that these reports do not just "re-present" the patient to the staff but actually serve as a substitute yet distorted sort of identity for the patient. The flesh-and-blood person is depersonalized as he or she is reduced to a short narrative of supposedly objective facts that is taken as everything that is relevant and important about that patient. This narrative, which is taken *as* the patient, actually is created, they explain, by the values of the medical establishment and traditions, which, though useful in some ways, can in other ways be damaging to the patient's interests. Thus literary theory and narratology are shown to be relevant to technical communication in the form of oral and written medical discourse.

Students could prepare an assignment for discussion or as a paper describing how the specific cases discussed by Poirier and Brauner relate to the feminist ethical concerns we have considered.

For still further research into ethics relating to information technology, a good place to start is Herman T. Tavani's "Information Technology, Social Values, and Ethical Responsibility: A Select Bibliography."

APPRAISAL, CASE FOUR: BURIED INFORMATION

Aristotle.   From an Aristotelian perspective emphasizing virtue and the deliberate cultivation of the habit of acting virtuously, one's senses of justice, of concern for the public good, and of honesty should all come into play in this case. It is inherently right and good, an Aristotelian would say, that any harm done to workers or to the public be correctly attributed and compensated. The company should take responsibility for the harm it has brought about. It is also right that workers be forewarned of the hazards they face so that they can make informed decisions about where and under what conditions they will work. The public too should be informed so that they are warned about potential dangers they face and can make informed public policy decisions affecting their lives. And it is right simply that the truth be made known and any mistaken assumptions about your plant and its operations be corrected.

To be sure, some ethical thinkers have pointed out another, common interpretation of Aristotle's ethics, namely one focusing on its concern for technical excellence and expediency. This is the pragmatic but ethically less noble side of Aristotle in an everyday sense. Steven B. Katz's article on Nazi human extermination technology is an example of this interpretation ("The Ethic of Expediency: Classical Rhetoric, Technology, and the Holocaust"). Dale Sullivan similarly has pointed out how basic principles of ethics from the classical period are often applied in ways that are shortsighted and potentially damaging to important social interests ("Political–Ethical Implications of Defining Technical

Communication as a Practice"). From this interpretation of Aristotle, your firm should be focused above all on maximizing efficiency and effectiveness as goals in themselves. Any health consequences to a few can be assimilated into the costs of the operation. Protecting the firm and minimizing its costs then would argue for not disclosing this information.

Kant.   From the Kantian perspective, the ethical determination is unequivocal. You should think of your ethical responsibilities in such a way that you would like others to behave toward you, too. If you were a worker in the plant, you would want anyone uncovering this research information to make it known to you so that you could protect your health. Likewise, if you were a citizen outside the plant, you would want to know about threats to your health and the environment. Because of the universal nature of this principle, it would make no difference who is the particular person involved or that person's circumstances (such as being the sole breadwinner of the family). The principle has an elemental force to it; it must be carried out regardless of secondary consequences.

Utilitarianism.   From a utilitarian perspective, the ethical determination would be so complicated that different people might arrive at completely different utilitarian conclusions. The basic principle is simple: Take the course that does the greatest good to the greatest number, balanced against competing bad effects. The calculation is problematic, though.

What are *all* the possible consequences? To whom? For what time frame—one day, one year, several generations? What weight is to be attached to each consequence? How do you weigh the effects of poverty or suffering?

Where do you draw the lines where consequences begin or end? If your blowing the whistle leads to the plant being shut down, that would be bad for the local population and economy. But it might be very good for populations and economies elsewhere, as might happen if your plant relocates to a foreign country or if the plant of a competitor in another country were to expand.

Feminist Perspective and Ethics of Care.   From a feminist perspective, impersonal entities and abstractions are perceived as threats to personal rights, welfare, and interests. If ethics is ultimately a matter of personal responsibilities, yet corporations are formed specifically to distance oneself from personal responsibility, then, many feminist thinkers contend, corporations are practically antithetical to what is usually considered ethics (at least insofar as this depersonalization is concerned). In addition, the perceived rights of a corporation to order its workers not to reveal certain information is understood by most feminist thinkers as actually less a right than an authoritarian expression of power that is innately unfair and inequitable—and unethical. The rights and health of the people concerned should be paramount.

From an ethics of care perspective, our ethical judgment should be guided by caring concern rather than by rigid adherence to abstract justice. Our caring concern for the health and welfare of the employees would urge that we reveal the information to workers and the public to help protect them. On the other hand, Noddings explains that caring about ourselves and those closest to us should temper our judgment. To engage in radically selfless behavior that jeopardizes our own welfare might therefore actually be unethical. We need to balance all the factors and potential consequences to all the people involved, including ourselves. If great harm would result to your family if you were to disclose, but little harm to only a few workers, then perhaps you should not disclose the information. On the other hand, you do not fully know the actual harm that the workers will suffer, as will all those related to them. Though this perspective makes ethical judgment more dependent on circumstances and feelings, it does not necessarily make them easier.

NB: This hypothetical case mirrors the actual case of Merrell Williams, a paralegal for a law firm that was working for the Brown & Williamson tobacco company. His job was to sort through many documents from the company's files. *Frontline* reports:

> Williams began to understand the depth of industry deception and began secretly copying the documents and taking them out of the building. By the time he was fired, he had copied 4,000 pages of documents. Among the documents was a memorandum by Brown and Williamson general counsel, Addison Yeaman (*Frontline* Web site, Settlement Case, p. 12, at http://www.pbs.org/wgbh/pages/frontline/shows/settlement/case/bergman.html).

The Yeaman memorandum is notorious for acknowledging that smoking is implicated in the occurrence of lung cancer. These documents proved vital to developing the criminal investigations against Brown & Williamson and other tobacco firms. Williams's ethical actions were particularly important because the industry has been so secretive, deliberately destroying documents or hiding them in obscure places.

APPRAISAL, CASE FIVE: WHISTLE-BLOWING

Case five does not involve the appraisal of a particular case but rather the exploration of whistle-blowing itself. It therefore does not lend itself to the sort of appraisal by each of the four perspectives as we have done for the other cases.

In the first part of this case, we consider whether ethics really expects anyone to behave "heroically" rather than ordinarily, and we consider the responsibilities of the public in relation to whistle-blowers. In the second part, we explore what one government agency involved in a great deal of scientific and technical research has done to clarify the issue of whistle-blowing. It has developed an explicit, detailed policy statement about when one should blow the whistle, how and why the whistle-blower should be protected against retaliation, and what institutional support should be offered to foster appropriate whistle-blowing and protect whistle-blowers from reprisals. In the third part, we see that there are times when one might have to blow the whistle on oneself. The potential consequences in the LeMessurier case were great, both to himself and to the stakeholders. The episode ends happily, however, and shows that sometimes acting ethically can be good business.

Though this does not deal with a specific instance the way the other cases do, we can still gain some insights into whistle-blowing abstractly, as a general principle, from the four ethical perspectives.

Aristotle.   The Aristotelian perspective would generally support whistle-blowing because it aims at doing good to stakeholders and at revealing the full truth of matters. In addition, the right thing to do is to protect the health and safety of those who depend on us to protect their interests. Justice in the sense of fairness and equity (rather than in the sense of rigid conformity to the law) would also urge us to blow the whistle when serious harm is likely. It is unjust for an organization to be allowed to act with impunity in ways that harm others.

Kant.   The Kantian perspective emphasizes duty and obligation. Though we might say that we have an obligation to follow the law no matter what, a broader interpretation of Kant would probably emphasize the dimension of universality. We should treat others as we would wish to be treated ourselves. Most of us would probably say that when serious danger or health hazards are involved, which is usually the case for real-world whistle-blowing, we would want and expect someone to blow the whistle.

Utilitarianism.   The utilitarian perspective, which emphasizes the greatest good for the greatest number would probably be supportive of whistle-blowing. Certainly within the narrow organizational context, whistle-blowing might bring little benefit and a great deal of damage to the organization. But the utilitarian perspective would also compel us to consider all the people involved, including all the stakeholders potentially affected. With this larger scope, clearly utilitarianism would support whistle-blowing, which almost always deals with issues of serious harm or risk to many people.

Feminist Perspective and Ethics of Care.   The feminist perspective for the most part does not approve of abstraction or of general rules to be applied universally to any and all instances and so might be disinclined to state an opinion on whistle-blowing in general. However, this perspective is highly supportive of exceptions to general rules on the basis of the unique particulars of any situation. Therefore, if we see whistle-blowing as the institutionalized acceptance of making exceptions, then the feminist perspective would certainly support whistle-blowing. It would be doubly supportive because whistle-blowing often occurs when health and safety are at issue.

APPRAISAL, CASE SIX: GOVERNMENT ETHICS

Government guidelines do not at first glance seem to be grounded in any particular ethical theory. These ethical guidelines do, however, clearly resonate with many of the ethical theories and principles we have been discussing. And this resonance is not just coincidental. This resonance should not be surprising because these same theories we have been discussing have given us the terms, concepts, and principles that we use in everyday talking about ethics and values. Thus governmental ethics often amounts to a common, accepted ethical theory that has reflected within it many formal ethical theories from throughout history.

Misuse of power, for instance, would be prohibited under Aristotelian ethics as being unjust and unfair; under Kantian ethics as violating the universal principle of equal treatment of all people; under feminist ethics as an expression of power for its own sake; and under an ethics of care perspective as not showing a caring concern for others. The same is true for conflict of interest, for which utilitarian ethics would also come into play. A utilitarian would argue that a conflict would impede the government from realizing the maximum possible social benefit from its activities.

Misuse of funds would be prohibited by utilitarian ethics, which would call for developing the most possible good (or return on investment) across all of society regardless of the particular persons involved. The appearance of impropriety, regardless of the reality, would fit well with Aristotelian virtue ethics as being "seemly," a credit to yourself and your society. Rules limiting certain kinds of postemployment activities would resonate with Aristotle's basic fairness and would resonate with a utilitarian desire for the taxpayers to get the most possible from their expenditures.

Impartiality, nondiscrimination, and the need for gender neutral and culturally sensitive language would strongly resonate with feminist ethics, which call for the remediation of past inequities and for treating all disadvantaged people the same as advantaged people. They would resonate with Kantian ethics, too, which calls for us to treat others universally as we would want to be treated ourselves. Kantian ethics also is based on "duty" or "obligation," precisely the language used in many governmental ethical guidelines.

Whether impartiality, nondiscrimination, or nonsexist language would resonate with Aristotelian ethics is debatable. In general, Aristotle repeatedly insists on justice, fairness, and a basic civic wholesomeness in one's relationships—the catch, though, is that for Aristotle this applied only to those advantaged or privileged people holding citizenship in Athens, and then only males (with very few exceptions). To all others, such as slaves, it was not only fair but entirely natural to act with less regard for fairness or justice. In our own times, we too have come to adopt the same basic Aristotelian frame of mind about justice but have broadened it to include everyone and exclude no one. Utilitarian ethics is also somewhat problematic. On the one hand, the government is committed to adhering to and enacting our constitutional principles. This root concern for principles regardless of costs would, of course, clash with the utilitarian concerns for maximizing benefits and reducing costs. School busing, for example, must be done regardless of the financial burden on the school district because the basic principle of nondiscrimination is more important than money. The same is true of special education programs to meet the needs of those requiring it.

From a feminist perspective, governmental ethics guidelines generally are a good idea but do not go far enough. The government has been on the forefront of efforts for inclusive language and multiculturalism, enacting the laws and rules that it itself has established. It has also been at the forefront of efforts to have environmental and other public consequences guide its policies and technical communications. It has been at the forefront of efforts to insist on plain language, standard formats, and high usability in its technical documents, too. This reflects a fundamental concern for the users and stakeholders of its documents and a high valuing of communication and relationship, which are to be commended from a feminist perspective.

On the other hand, the government can also be granted exemptions that no other entities are entitled to, especially on the basis of national security. Some of the worst sites of environmental pollution in our country are government installations for researching, testing, or developing nuclear materials or weapons. The justification given for such exemptions is the worthiness of the goals—the security and continued existence of the country itself. These exemptions therefore have a utilitarian rationale. In such cases, the government can show a chilling disregard for the care of many people and for the environment, such as revealed about the secret experimentation on the public of the effects of radiation (see ACHRE reports). This would clearly be deplored from a feminist perspective as well as from an ethics of care perspective.

REFERENCES

Cushman, Reid. "Information and Medical Ethics: Protecting Patient Privacy." *IEEE Technology and Society Magazine* Fall 1996: 32–39.

Del Tredici, Robert. *At Work in the Fields of the Bomb.* New York: Harper and Row, 1987.

Katz, Steven B. "The Ethic of Expediency: Classical Rhetoric, Technology, and the Holocaust." *College English* 54/3 (1992): 255–75.

Markel, Mike. "A Basic Unit on Ethics for Technical Communicators." *Journal of Technical Writing and Communication* 21/4 (1991): 327–50.

Martin, Mike M. "Whistle-blowing: Professionalism, Personal Life, and Shared Responsibility for Safety in Engineering." *Business and Professional Ethics Journal* 11/2: 21–39.

Morgenstern, Joe. "The Fifty-Nine Story Crisis." *New Yorker* May 29, 1995: 45–53.

Poirier, Suzanne, and Daniel J. Brauner. "Ethics and the Daily Language of Medical Discourse." *Hastings Center Report* August/September 1988: 5–10.

Sullivan, Dale. "Political–Ethical Implications of Defining Technical Communication as a Practice." *JAC: Journal of Advanced Composition* 10 (1990): 375–86.

Tavani, Herman T. "Information Technology, Social Values, and Ethical Responsibility: A Select Bibliography." *IEEE Technology and Society Magazine* Summer 1998: 26–39.

Wujek, Joseph H. "Must Engineers Behave Heroically?" *IEEE Technology and Society Magazine* Spring 1996: 3.

INDEX